iSpeak Cloud: Embracing Digital Transformation

First print edition, May 2016

ISBN: 978-0-9846757-5-3

Visit us on the web: www.ispeakcloud.com

Credits

Graphic Design and Artwork: Paul Daniels and Wire Stone, LLC.

Producers: Michael Procopio, Elaine Korn and Muneer Mubashir

Assistant Producers: Sheila Mangione and Jeanne Morain

Editor: Mangione, Payne and Associates

Dedication

The iSpeak Cloud series is dedicated to the men and women who have burned the midnight oil to identify and resolve gaps in people, process and technology over the last two decades. They have embraced change and forged new ground from the client to the cloud and, in their enterprises, they are the drivers of digital transformation.

It is dedicated also to my family and friends for their love, support and sacrifices over the years. In particular, I include late nephew Steven Smith, who may be gone but will never be forgotten, his brother Chris, a true hero to the community, our country and our family, and my surviving nephew Ray Morain for his courage and strength on his long journey to recovery.

Heroes

A special thank you to Hewlett Packard Enterprise for sponsoring and supporting the independent research contained in this book. The contributors come from a wide variety of public- and private-sector entities and serve in diverse roles and disciplines across industry, business and technology. iSpeak Cloud recognizes and extends a special thank you to everyone who took the time to interview and/or provide feedback for this book, including:

Lenin Aboagye
David Angradi
Chris Armstrong
Srikanth Balusani
Matt Bausch
Jim Berry
Charlie Betz
James Burdick
Paul Burnham
Claudiu Burdelean
Chris Cason
Greg Chamack
Steve Chambers
Bryan Diehl
Craig Faruncheck
Ryan Fee
Jocelyn Goldfien
Ben Gordon
Linda Kavanagh
Tom Klemzak
David Langlais
Louis Lendi
Nate Lenoir
Jennifer Lewis
Andi Lim
Michael Mattal
Evan Maxey
Muneer Mubashir
Maria Oliveria
Paul Peisner
Michael Procopio
James Rhodes
Tyler Rohrer
Andy Santosa
Bob Small
Pat Spica
Andy Santosa
Tiffany To
Ty Tobin

Change Agents

There are a few people I want to highlight not only for their contributions to the book but also their phenomenal successes in promoting positive organizational and technology change with proven results.

Gary Acromite, Chief Information and Logistics Officer, Gavilon Group

Aaron Amendolia, Vice President of Service Delivery, National Football League

Kia Behnia, CEO, PowWow Mobile

Mark Bodman, IT4IT Strategist

Winston Bumpus, President of Desktop Management Task Force

Brian Cinque, Cloud Service Leader, General Electric

Don Cox, Chief Information Security Officer

Marios Daminades, Partner at E&Y and Board Advisor, ISACA

Emerald de Leuuw, CEO of Eurocomply

Malcolm Fry, Industry Luminary and Author

Shirley Gao, Chief Information Officer, Young's Market Company

Lorenzo Hines, Portfolio Leader, Citi Cards

Chad Jones, Chief Executive Officer, Virtual Strategies

Andi Mann, Chief Technology Advocate, Splunk

Dr. Victoria Molfese, Chancellor Professor, University of Nebraska and Associate Editor Developmental Neuropsychology

Jim Noble, Former Chief Information Officer, General Motors, AOL and Merrill Lynch

Veronica O'Shea, Former Executive Vice President, eBay, Senior Vice President, SAP, and General Vice President and General Manager at Oracle

Alex Ryals, Vice President of Internet of Things, Cloud and Security, Avnet

Cindy Schumacher, Former Chief Information Officer, Cognizant and Expedia

Oren Taylor, Director, CDG Europe

William Velez, Chief Information Officer, Global Financial Institution

Ruth Veloria, Executive Dean, School of Business, University of Phoenix

About the Author

Jeanne Morain is the principal researcher and consulting strategist at iSpeak Cloud. She has held various executive roles in strategy and product management with the Apollo Group, Flexera Software, VMware (Thinstall) and BMC Software (Marimba). Jeanne currently advises startups and large enterprises on implementing new products and strategies to enable excellence in the digital economy.

Jeanne has two decades of experience in systems management, virtualization and cloud computing and has participated in the implementation of solutions for millions of users across Fortune 2000 companies. She has won numerous awards for her work, including the prestigious International Association of IT Asset Manager's Fellow Recipient in 2016 for her work in business service management, Lifetime Member Award in the areas of business service management, universal clients (also known as virtual desktop infrastructure), dynamic data center and virtualization. She is an author and coauthor of books on BSM, virtualization and cloud computing.

Jeanne is best known for her customer-/partner-centric approach to research and solutions. Based on her involvement in dozens of enterprise software implementations and the lessons learned gathered through interviews with 180+ business and technology leaders, she has developed templates, processes and prescriptive guidance that help enterprises to not only accelerate the transition to cloud but also transform themselves to compete in the digital age. She is a noted industry speaker at VMworld, Interop, CloudSlam, IAITAM, CXO events and other user conferences. She has written blogs as well as articles for trade publications. Jeanne holds a Master's degree from Southern Illinois University and certification in ITIL. www.ispeakcloud.com, twitter @JeanneMorain.

About Hewlett Packard Enterprise

Hewlett Packard Enterprise is an industry-leading technology company that enables customers to go further, faster. With the industry's most comprehensive portfolio, spanning the cloud to the data center to workplace applications, our technology and services help customers around the world make IT more efficient, more productive and more secure.

A Technology Leader's Perspective:

Alex Ryals
Vice President of Solutions Development
Avnet, Inc.

Embracing Transformation before IoT Becomes Mainstream

Many enterprises are hybrid by happenstance, not by choice. They didn't implement hybrid cloud solutions as part of an overarching strategic initiative. Just as they struggled in the past with virtual sprawl and stall, today they are feeling the growing pains of cloud sprawl.

Over the last two years, Jeanne and I have been teaching workshops on how to consolidate and create a cohesive strategy to cross the proverbial cloud chasm. The concepts in the five-stage process in her *iSpeak Cloud* books are relevant whether your organization is a small one or a large one. With cloud-based solutions moving into the mainstream and newer disruptive technologies emerging, it's more important than ever for leadership from IT and the business to unify their strategies, eliminate silos and automate manual processes.

Internet of Things (IoT) will bring an even greater level of change and disruption as more devices and threat points connect in the cloud. Highly publicized breaches in this area have taught us that unsecured cloud portals not only lead to issues with IoT devices but also with adjacent company systems to which those devices are connected. The security services mentioned in this book—microservices that scrub connecting services for vulnerabilities and compliance—will be imperative as more of these devices are connected to enterprise and shared third-party networks.

The core principles of incorporating business, security and regulatory compliance into initial planning and design are critical for building a solid foundation to quickly onboard cloud services while reducing risks. To Jeanne's point, the industry is undergoing a major transformation across the layers in the stack. Everyone across the ecosystem from the consumers to the providers of the technology solutions need to adapt and embrace the change.

A Business Leader's Perspective

Veronica O'Shea

Former Executive Vice President, eBay, Senior Vice President, SAP, General Vice President and General Manager, Oracle

New Realities in the Digital Age

The digital natives described in this book are transforming the way companies must market, sell and pivot to customers. Digital consumers are connected and empowered like never before. Technology enables them to buy virtually anything instantly, from any device, and take delivery anywhere. In the digital age, consumers are in control and successful companies must consistently and systematically meet their needs.

Providing customers with an omnichannel experience—that is, interacting across every channel the consumer wants to use—is more of a requirement than an advantage in today's market. Consumers have many options at their fingertips and they know it. They won't hesitate to jump to a competitor who does a better job of anticipating and responding to their needs. Business leaders, including chief information officers (CIOs), have to embrace that new reality and adapt quickly to deliver the technology-based solutions required for survival.

This book's approach to digital transformation sketches out a vivid roadmap for success. Gone are the days of IT being either a segregated or integrated provider of services. In the digital enterprise, technology is the very essence of business success. Business leaders know their areas very well and should be empowered to implement solutions that enable them to adapt to and embrace change to meet the needs of their customers. To survive and thrive in this new world, business leaders need the framework and guardrails this book describes to transform their enterprises without introducing risk or driving up costs.

Implementing the business discipline of traditional product management to fit the service delivery needs of customers has been Jeanne Morain's passion. And her passion has driven innovation and enabled her clients to evolve. The wisdom in this book comes from a depth and breadth of experience along with the ability to listen and learn. It is a must read for any business leader trying to cut through the confusion in an age of information overload.

Preface

"Digital transformation and enterprise hybrid cloud strategy are the chicken and egg. Hybrid cloud expedites digital transformation and digital transformation accelerates the adoption of hybrid cloud and expands its reach."

—Shirley Gao
CIO, Young's Market Company

iSpeak™ Cloud: Embracing Digital Transformation is the second book in the iSpeak Cloud series. It provides prescriptive guidance regarding tools and implementation strategies for the framework and concepts introduced in the first book, *iSpeak™ Cloud: Crossing the Cloud Chasm.*

The term *cloud* has a lot of hype associated with it. People define cloud in many different ways and that has resulted in confusion about what cloud is and isn't. The iSpeak Cloud books use the National Institute of Standards and Technology definition which states that a cloud solution must, at a minimum, possess five characteristics: on-demand self service, broad network access, resource pooling, rapid elasticity and measurability.[1]

iSpeak™ Cloud: Embracing Digital Transformation subscribes to the idea that no one solution fits every digital transformation. Transformation needs vary depending on customers, employees, processes and the current technology state. The book provides insight into the digital era, including its evolution from traditional IT and business service management. Moreover, it helps enterprises build a framework for implementing the cloud solutions required for digital transformation. The templates, tools and guidance are based on real-world implementation best practices within Fortune 2000 companies.

A book coauthored by Jeanne Morain, *Visible Ops Private Cloud*,[2] outlined the steps for creating a solid foundation on which to build a private cloud and position IT to shift from the role of service provider to the role of service broker. Many of the interviewees for that book were just beginning to embrace cloud computing and were establishing policies to guide this shift. They viewed private cloud as a first step.

---

1. *The NIST Definition of Cloud Computing*, National Institute of Standards and Technology, 2011. Available from http://www.nist.gov/itl/cloud/index.com.
2. *Visible Ops Private Cloud: From Virtualization to Cloud in 4 Practical Steps*, Andi Mann, Kurt Milne and Jeanne Morain, IT Process Institute, 2011. Available from www.itpi.org.

In a book she later authored, *Client4Cloud*,[3] she expanded on the impact of the shift to the cloud and the overall technology adoption by digital natives who were beginning to come of age. The shift to a user-centric perspective helped drive a corresponding transition to hybrid cloud architecture. That brought additional risks as well as new regulations and corresponding costs to the enterprise.

In *iSpeak™ Cloud: Crossing the Cloud Chasm*,[4] Jeanne expanded the prescriptive guidance to creating a higher-level process for digital transformation to bridge the gap between where the business needs to be and IT's ability to deliver solutions across the various clouds, devices and channels to reach that destination. To bring the concepts to life, she wrote the book in the format of a script for a play. The primary objective was to provide real-world tools, templates and solutions for elevating the business perspective of the IT organization from a service broker to a trusted advisor.

In writing these four books Jeanne conducted more than 180 interviews with business leaders, analysts, vendors and enterprise IT professionals. She also drew on her own experience, which involved participating in similar major shifts, such as business service management (BSM), regulatory compliance, virtualization and user-centric universal computing. All this experience gave her a broad and deep understanding of the benefits and challenges of the digital transformation.

The intent of this book is to provide a governance framework that will guide IT organizations in successfully implementing hybrid cloud solutions that drive digital transformation.

Figure 1. Jeanne Morain books by year

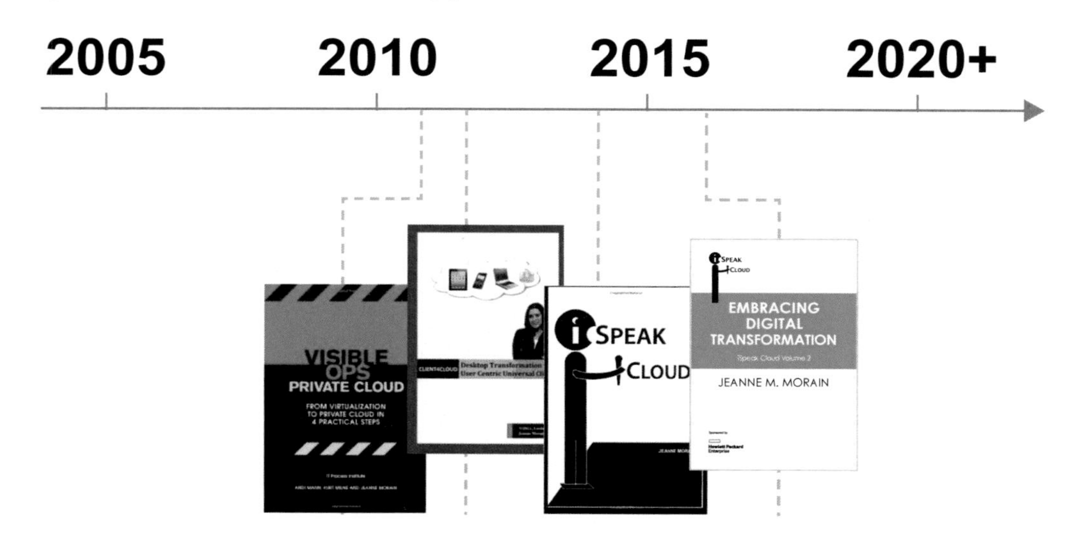

---

3. *Client4Cloud: Desktop Transformation to Dynamic Universal Clients*, Jeanne Morain, Coolmody, LLC, 2011. Available from http://www.amazon.com/s/ref=nb_sb_noss?url=search-alias%3Dstripbooks&field-keywords=Jeanne+Morain.

4. *iSpeak Cloud: Crossing the Cloud Chasm*, Jeanne Morain, iSpeak Cloud, LLC, 2014. Available from http://www.amazon.com/s/ref=nb_sb_noss?url=search-alias%3Dstripbooks&field-keywords=Jeanne+Morain.

Table of Contents

INTRODUCTION

"The transformation from analog to digital happened years ago. Today's market disruption is driven by the way users in IT as well as in the business expect to consume and use digital services. Hybrid cloud is at the epicenter."

—Jim Noble
Former CIO of General Motors, AOL and Merrill Lynch

The use of technology in our everyday lives has become ubiquitous. Internet usage increased 832 percent between 2000 and 2015—bringing the number of internet users to more than three billion.[5] Today's consumers turn to PCs, laptops, smartphones and tablets to shop, bank, make travel arrangements, interact with friends and learn new things.

The digital revolution is driven by the same factors that sparked the transportation revolution: The automation of previously manual data center operations processes is bringing down the cost of technology and IT services, resulting in growing consumer acceptance of and demand for digital services.

IT needs to be proactive, leading the enterprise through an essential digital transformation. Many IT organizations, however, are struggling to keep up with the rapid change that going digital requires. Cloud technologies are at the heart of the transformation and IT must embrace cloud to increase agility, drive efficiency and enable business innovation.

The purpose of this book is to help IT organizations successfully navigate the road to cloud. By spearheading the digital transformation, IT can become a partner to the business and a

5. *Internet World Stats: Usage and Population Statistics, http://www.internetworldstats.com/stats.htm.*

true enabler of business value. After reading this book you will:

- Understand the challenges cloud presents as well as the key people, process and technology competencies needed for a successful digital transformation.
- Understand how to establish a cloud center of excellence (CCoE) and supporting competency centers to create a governance framework and an execution plan for the journey to cloud.
- Be empowered with a five-step approach to implementing a hybrid cloud strategy that facilitates digital transformation.
- Understand which IT and business processes to automate, which to facilitate and which to eliminate.

Parallels with the Transportation Revolution

It takes more than technology and engineering to drive change and set off seismic market shifts. While the invention of the automobile can be traced back to 1672,[6] it was not until Henry Ford built the Model T assembly line that cars became available to the masses. By standardizing processes and automating manual tasks, Ford's assembly line reduced vehicle production time from 12 to 2½ hours,[7] slashing production costs and making cars more affordable. As the number of cars on the road went up and prices went down, adoption by consumers grew. In 2010, the number of automobiles worldwide surpassed one billion, and it's difficult to imagine a world without this essential means of transportation.[8] The digital revolution is following a similar path.

It took almost **250 years** for automobiles to become mainstream.

6. https://en.wikipedia.org/wiki/History_of_the_automobile#Early_automobiles.
7. http://www.history.com/this-day-in-history/fords-assembly-line-starts-rolling.
8. Huffington Post, 8/23/11, "Number of Cars Worldwide surpasses 1 Billion," http://www.huffingtonpost.ca/2011/08/23/car-population_n_934291.html.

Where We Are Today

Rapidly emerging cloud technologies are enabling the digital revolution. The coming together of technologies including virtualization on the mainframe coupled with proven results has paved the way for the adoption of cloud solutions. Open source and standards such as the Distributed Management Task Force's (DTMF) Open Virtualization Format have pushed vendors and developers to commoditize cloud consumption. In parallel, companies such as Netflix have proven to businesses and consumers that successful ideas can be supported, scaled and expanded in the cloud.

The adoption of *as-a-service* in the cloud has hit critical mass. And, as with the adoption of mobile-cellular technology, cloud consumption with consumerized applications has become an integral part of day-to-day business operations.

Figure 2. **Clash in cloud**

The convergence of digital consumers (people), consumerization/standardization (process) and disruptive technology has fueled the perfect storm fueling digital transformation.

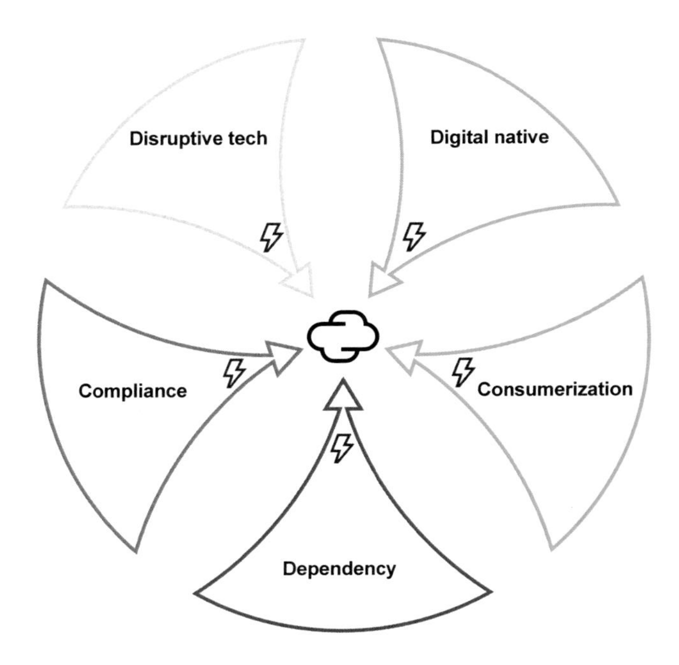

The Need to Reinvent IT in Traditional Enterprises

Organizations such as Google, Facebook, Netflix, Uber and Airbnb were born as digital enterprises. They were founded in response to the growing demand for digital services—and the digital nature of their business models poses a significant competitive threat to traditional businesses born before the digital revolution. In response, savvy CIOs in traditional enterprises are reinventing their IT organizations to be an integral part of the business. Below are some examples of unique, compelling solutions that drive enterprise competitiveness:

- Agricultural industry leaders are leveraging DevOps and hybrid cloud and reallocating resources from IT to the business to increase velocity not only within IT but across the logistics supply chain to help farmers get products to market efficiently and profitably.
- Health information exchanges are enabling physicians and supporting hospital networks to differentiate their care and comply with regulatory mandates.
- Cellular providers have created music stores that are self healing and customer focused.
- Financial institutions have introduced competitive differentiators such as mobile deposit and payment into their online banking services.

Embracing the Digital Native

In the interviews conducted for the iSpeak Cloud books, digital transformation has been a common theme. In the most recent interviews, however, it's clear that a new factor is shaping the digital economy: the coming of age of digital natives. These are individuals born during or after the introduction of digital technologies—starting in the early 1990s. *Digital natives* have never known a world without computers, the internet or cell phones. According to top neuroscientists, the digital native's world is unprecedented.

To the digital native, using technology is as automatic as walking or talking. And while consumers of all ages are taking advantage of exciting new digital services, the digital natives' relationship with technology differs markedly from that of previous generations. The digital natives' consumer, work and social behaviors are formed by their relationship with technology and are not easily modified or controlled. The generational divide between digital natives and baby boomers is greater than any other in the history of the human race. Digital natives consider the processes and technologies of companies that existed before the cloud era—which began about the time Amazon introduced its Amazon Web Services (AWS)—to be outdated and constraining.

In planning a cloud strategy, IT organizations must keep digital natives in mind, not only as customers but also as employees. Interviewees from traditional companies report that retaining digital native (millennial) talent is one of the biggest challenges they face. Their workforces tend to be 80 percent traditional employees and 20 percent digital natives. In contrast, companies created in the digital era, with their cloud-by-design model, report just the opposite. Nearly 80 percent of their workforce are digital natives and only 20 percent belong to previous generations.

"The human brain is highly adaptable and will adapt to environmental variables. For example, experiences with digital technology during childhood and throughout adolescence have transformed the way digital natives think, act and adapt to their environment. These experiences in the use of digital technology as tools or strategies for exploring the environment and problem solving work to build and strengthen neural networks in the brain. Further, because of the multimodal capabilities of computers, users with different cognitive and information processing styles can be accommodated."

— Dr. Victoria Molfese
Chancellor Professor, University of Nebraska and
Associate Editor Developmental Neuropsychology

Digital natives as customers and employees represent a huge opportunity for enterprises working toward digital transformation. These people are the new generation of customers. Consequently, a savvy digital approach enables an enterprise to attract and retain these consumers and build long-term relationships with them. From an employee standpoint, digital natives are positioned to help senior management understand the interests and needs of this expanding consumer base so the enterprise can create compelling solutions targeted to them.

Understanding the Cloud Chasm

Legacy IT departments were built to develop, mandate and control processes. The IT model at the core of their operations simply doesn't meet the needs of the digital enterprise. Although most IT organizations are attempting to implement cloud, there is a gap between where the enterprise believes it needs to be with respect to cloud-based technology and IT's ability to deliver solutions across various clouds, devices and users while balancing business requirements, compliance and resource availability.

This gap—or cloud chasm—is widening in traditional IT shops because employees, not only from the business side but also from the IT side, are rapidly adopting solutions outside of the control, scope and understanding of IT—increasing the footprint of a situation known as shadow IT.

Many of the IT professionals interviewed for the first iSpeak Cloud book indicated they had been caught off guard by what has become an epidemic facing their companies. Business and IT users, especially digital natives, are frustrated with IT's inability to deliver the services they need in a fast, reliable way. They don't want to deal with older processes and hierarchical layers. These users resist processes that were developed for a previous era. A number of factors are motivating them to circumvent IT, including:

- Ease and accessibility of purchasing cloud-based tools
- Tech-savvy employees looking to improve their work/life balance
- Growing number of regulations and standards around technology for audit and compliance
- Shortage of IT resources, which results in implementation delays
- Brownfield solutions that need to be augmented, integrated or replaced
- Lack of a common language between IT and the business, leading to miscommunication and adversarial relationships
- Lack of alignment between objectives of business leaders and IT leaders
- Software-as-a-service (SaaS) solution providers that have oversold benefits and undersold total costs

> "Everyone in the company must understand that technology is a critical component of business. Technology leaders must embrace change to build a partnership of trust and have a seat at the table. Business leaders should work with IT to automate legacy processes that hinder progress."
>
> — Ruth Veloria
> Executive Dean, University of Phoenix Business School

More and more employees are going around IT and deploying public cloud solutions on their own, which is increasing the size of the shadow IT environment. Without IT oversight, shadow IT instances inhibit the enterprise's ability to protect corporate data and intellectual property (IP), contain costs and ensure regulatory and industry compliance. As a result, the majority of companies do not have a firm handle on the various types of applications, services, and solutions that are deployed. Their environments end up looking like a multiheaded hybrid hyrda with a mix of legacy and cloud based applications making up the service. Because of the lack of visibility and control - the IT team is in a constant battle trying to address security, compliance and costs risks as they come up.

The hardest part of crossing the chasm is breaking down the silos that separate not only IT and the business but that also isolate individual IT groups and prevent collaboration. Companies that have had to struggle the hardest to address internal politics report higher numbers of shadow IT applications and an inability to execute.

Successful IT leaders pull all the stakeholders together in embracing change. They evaluate shadow IT in their organizations and find ways to enable users instead of inhibiting them.

Figure 3 below represents the current chasm that exists between leadership across technology team and their peers.

Figure 3. **Cloud chasm**

The Importance of Hybrid Cloud

Most companies, both traditional and digital by design, already have or are planning hybrid cloud implementations. A January 2016 *Wall Street Journal* blog cites an International Data Corporation (IDC) report that estimates worldwide spending on public cloud services will grow by a 19.4 percent compound annual rate over the next four years, from $70 billion in 2015 to $141 billion in 2019. In addition, the article states that cloud applications will double for IT over the next year.[9]

All interviewees for this book agreed that hybrid cloud, which combines an enterprise's internal cloud environment with the public cloud, is an essential ingredient of a digital transformation strategy. Hybrid cloud gives enterprises the freedom to assemble the optimum mix of private and public cloud services to satisfy the unique needs of their businesses.

In the last few years, the appetite for interactions across multiple digital channels (omnichannel interactions) is introducing significant challenges. IDC's 3rd Platform vision holds that businesses need to embrace cloud, mobile, social and big data technologies to reinvent traditional models for both the business and information technology.[10] Although DevOps provides an effective methodology to accelerate the rate of change in older frameworks such as IT Infrastructure Library (ITIL), it cannot bridge the gap in the traditional enterprise between technology consumers and providers.

The hybrid hydra shown in Figure 4 represents the majority of IT environments today. Services are made up of a conglomeration of solutions across multiple clouds and legacy environments.

Figure 4. Hybrid hydra

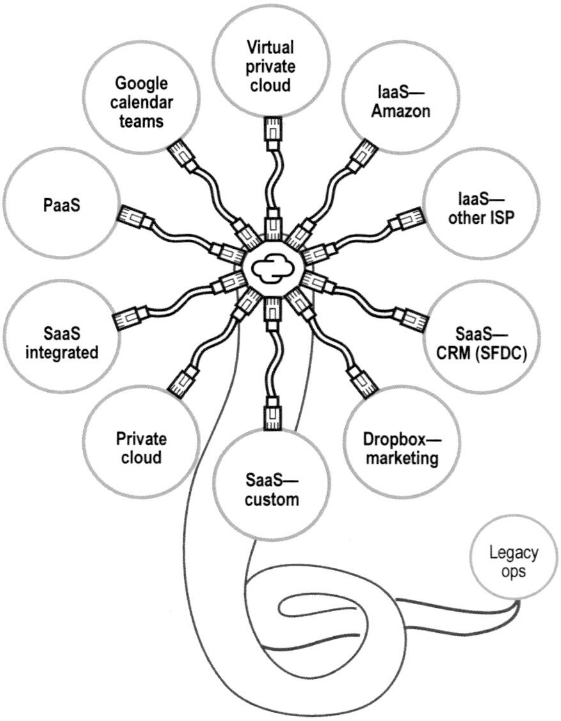

---

9. "Enterprise Cloud Adoption Gains Steam: Report," *The Wall Street Journal*, January 2016. http://blogs.wsj.com/cio/2016/01/21/enterprise-cloud-adoption-gains-steam-report/

10. IDC, http://www.idc.com/prodserv/3rd-platform/ 2016.

The Pressure to Go Digital

With the digital revolution shifting into high gear, traditional organizations in the public and private sectors must reinvent how they do business to compete. They are grappling with how to best handle consumer demand for omnichannel interactions that allow people to engage via the web, social media, text, chat and more. That means not only transforming their legacy IT infrastructures to modern cloud architectures, but also reinventing their organizational structures to make the most of this technology transformation.

The pressure to go digital is intense, but successful enterprises won't rush through it. Instead, they will follow sound business practices so that they don't expose their organizations to business risk. They will carefully think through the process to ensure compliance in today's highly regulated global economy. Traditional IT must reexamine conventional thinking and decide what is worth keeping versus what technologies and processes need to be left behind. And as cloud usage expands, they will establish a clear framework for cloud governance and management that maintains security, compliance and cost effectiveness.

iSpeak Cloud: Embracing Digital Transformation will empower you to manage the digital transformation in your enterprise. It describes best practices for enhancing policies, skills and strategy and embracing the shift to cloud while minimizing risk and keeping costs in check.

The key to the successful planning and execution of the digital transformation and the primary focus of this book is the cloud center of excellence (CCoE). The CCoE is a cross-functional team that provides leadership, best practices, guidelines and support to enable the implementation of a cloud service platform that drives business success in the digital economy.

As with other paradigm shifts, the winners will be the enterprises that embrace the change. Are you ready?

DIGITAL TRANSFORMATION, AN OVERVIEW

"Technology can build a great infrastructure and magnificent machinery but whether or not it is successful boils down to how it's being consumed by the applications and users. Essentially the consumer is the future administrator of the data center. It's important to build an abstraction layer that automates as much of the shared processes as possible to facilitate success."

—Winston Bumpus
President, Distributed Management Task Force

Traditional IT service delivery models aren't geared toward supporting the level of innovation required in the digital economy. IT must evolve its role from one of delivering services such as email or data entry to the business to the role of partner to the business.

Figure 5. **Four levels of digital transformation maturity.**

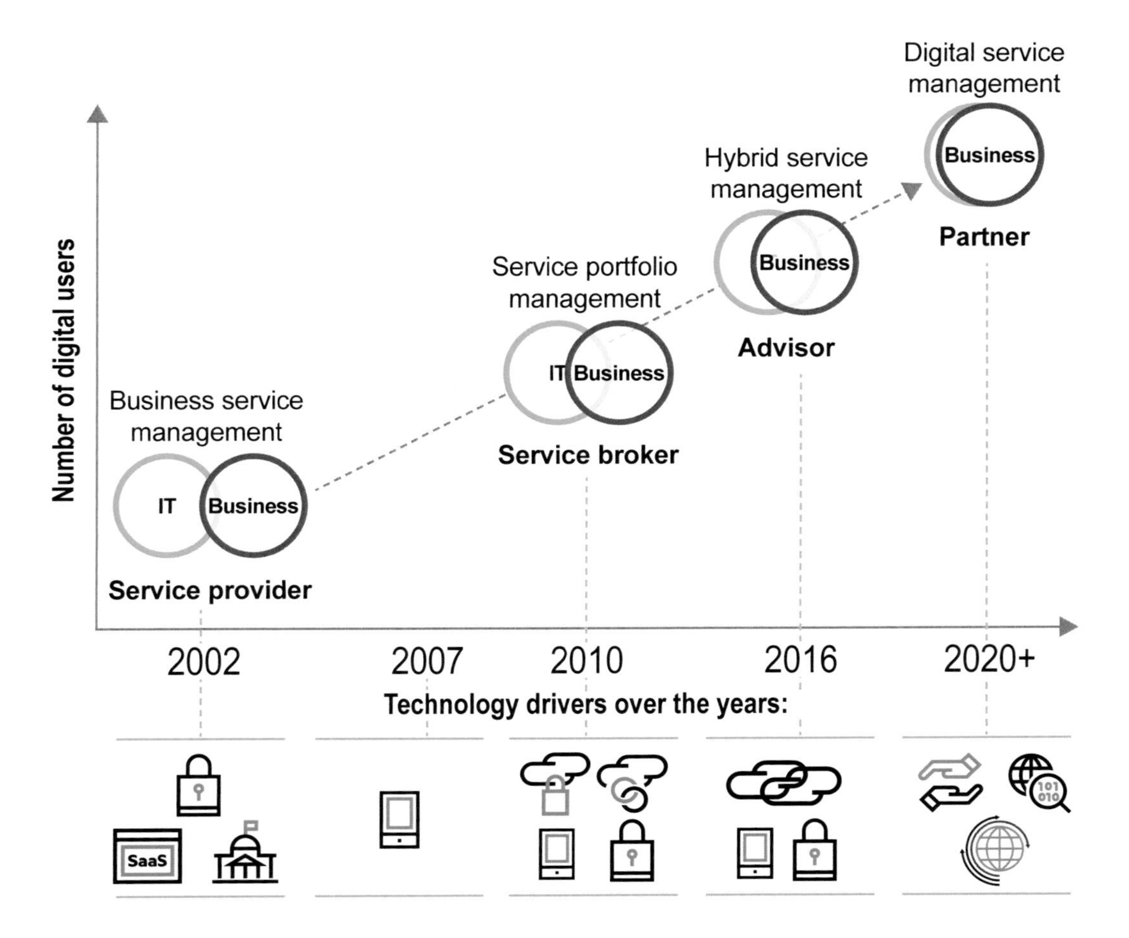

The companies interviewed for this book fell into four distinct levels of digital evolution, which are depicted in Figure 5:

1. **Service provider**—These enterprises are characterized by the traditional role of business analysts, layers of process approvals and a disconnect in communication between IT and the business. Enterprises at this level had the highest incidence of shadow IT and the lowest levels of innovation velocity. Employees from both the business and IT expressed frustration with the inertia created by archaic process layers that prevented them from moving at digital speeds.
2. **Service broker**—Enterprises at this level have shifted to a service portfolio view. They are still traditional in their thought processes and heavily focused on ITIL processes to balance the portfolio. They are beginning to combine private and public cloud. Quite a bit of shadow IT may be present because the IT staff cannot keep up with the business demands for innovation.

3. **Trusted advisor**—In these enterprises, the business consults IT on major purchases because IT has earned the trust of the business side of the house. The CIO promotes an integrated team approach and is beginning to facilitate discussions with the business around cloud technologies and how they can be applied to expand the business. These companies may or may not have a DevOps strategy. However, they enjoy high innovation velocity through process automation.

4. **Partner**—These enterprises understand that "technology is the business." They have already hired a chief digital officer (CDO) or the chief information officer (CIO) has assumed this role. The digital culture is driven from the top down. Although DevOps is used widely in these enterprises, interviews for this book indicate that not all digital-by-design enterprises use DevOps. A few of them reported that the processes they have created are just as effective for meeting service level agreements (SLAs).

The age, experience and technical maturity level of both its customers and its employees affect how quickly a company can move from one level to another. For example, a music service provider that employs and sells primarily to millennials can adapt quickly to a digital-by-design service offering from its traditional manufacturing process. However, a large distributor of music CDs with a large reseller and employee base composed of older consumers and employees may opt to start out as a service broker and move to a trusted advisor. This will allow IT to iterate not only its offerings but also how the staff services internal and external customers.

Regulations also play a role in determining how quickly or even if a company can move from one level to the next. The more global the company, the bigger the compliance challenge. For example, retailers selling goods worldwide may have to keep certain systems on premise to comply with privacy regulations across the countries in which they sell. The changing regulatory landscape across countries causes many senior leaders to hesitate to sign up for services that are not located in their primary country. In some cases, uncertain economic and political conditions make it more conducive for companies to host data and resources in countries that have more stable political conditions.

The ability to attract and retain digital native talent increases as the evolution progresses. Enterprises that fall within the traditional service provider or service broker level struggle to retain millennial talent for a long period unless they create some form of incubator program. A number of top performers are already implementing programs that create an environment to foster innovation by insulating groups of digital natives from the rest of the workforce to reduce their exposure to processes they might find onerous until they acclimate.

One Size Does Not Fit All

The level of maturity and phase of adoption vary from one enterprise to another. Top performers do not subscribe to a single approach. They create their own recipe for success by balancing the business they are in with prescriptive guidance across multiple disciplines. There are three critical factors to consider when creating your transformation strategy:

- Your core business offering
- Age, experience and level of maturity of your employees and customers
- Regulations and risk not only to the technology but also to company brand

Companies such as Google and Facebook were born on the internet. They don't have to become digital; they are digital by design. If the internet ceased to exist, they would be out of business. Technology is their business. Their users are connected and assumed to be at least somewhat technical.In companies for which the primary business isn't digital in nature, IT may not progress all the way to the partner level of the digital transformation spectrum due to the role IT plays in the business. For example, in healthcare, if the electronic medical records system goes down, a doctor can still save lives. If an agricultural system is unavailable to monitor moisture and soil conditions, the farmer can still tend his crops. If the National Football League (NFL) website goes down, Super Bowl players can still take the field and the game will go on. In these companies, IT may move some applications to the cloud and adopt some elements of a digital-by-design model but may strive only to reach the trusted advisor level of the spectrum.

How Top Performers Approach Digital Transformation

The top-performing companies interviewed for this book spent more time in planning and less in delivery and execution than their lower-performing counterparts. As one interviewee put it, "To truly go fast you have to slow down and create a plan first, instead of just running ahead."

"To truly go fast you have to slow down and create a plan first, instead of just running ahead."

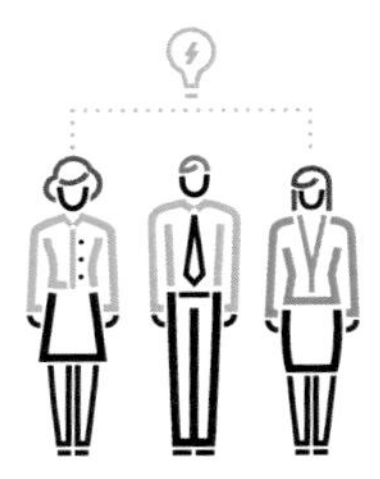

The CIOs of companies that are achieving success in the digital economy exhibit several common traits:

- **Facilitation** – They offer a seat at the table for key stakeholders across business, technology and governance.
- **Guardrails instead of roadblocks** – Instead of creating obstacles that prevent business users from getting what they need, they create and enforce policies and governance models that guide users' choices.
- **Emphasis on planning** – They make decisions and develop strategy based on business cases that reflect return on equity across the lifespan of equivalent applications for the company.
- **Accountability** – They hold both business and technology teams accountable for plans around project profits or savings from the implementation of a solution or service.
- **Automation** – Top CIOs and chief digital officers (CDOs) work with the business to identify areas of automation to reduce inertia and increase velocity, agility and quality.
- **Willingness to embrace change** – Rarely are they caught in the hype cycle, nor are they bleeding-edge adopters. However, they are often in the first 15 to 20 percent of enterprises to adopt compelling technologies, processes and methodologies.
- **Proactive and preventative approach** – Security, compliance and measurement are part of planning and design. Senior leaders sleep well at night knowing that although breaches may occur, the impact to the business will be minimal due to preventative measures.

Top-performing enterprises know the market space they serve and the technologies in that space—particularly technologies that are as far reaching and impactful as cloud. Lower-performing companies, on the other hand, often confuse cloud with server virtualization or SaaS applications hosted at someone else's site.

Figure 6. **Digital transformation approach**

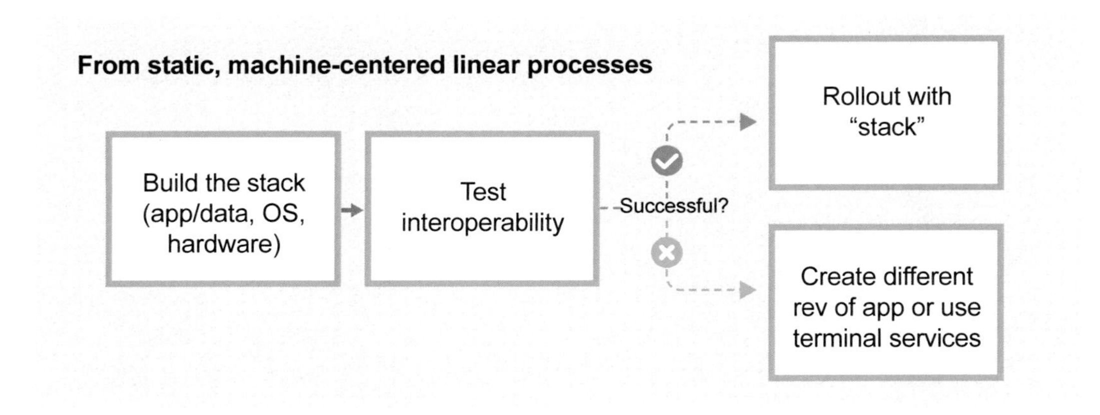

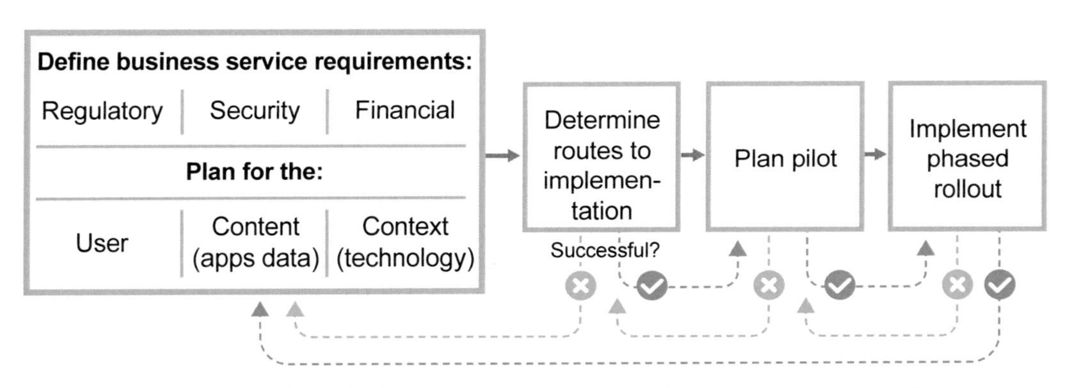

Companies are shifting from a static process to one that is multi modal taking into account all aspects of the business service with iterative loops for continuous improvement, integration, and delivery.

Five Phases of Digital Transformation

Although every enterprise transformation will be unique depending on goals, current maturity level and culture, top-performing companies go through five common phases to transform current processes, skills and tools and embrace the user-centric paradigm shift across the various layers in the stack from the infrastructure to the application. Figure 7 summarizes the phases and the goals of each one. The remainder of this book dives deeper into each phase to illustrate what transformation success looks like.

Figure 7. **The five phases top performers use to achieve digital transformation.**

Phase	Goal
1. Cloud Command and Control	Create an executive-level governance team—a cloud center of excellence—of cross-functional business leaders to define the transformation strategy and create rules and policies to facilitate transformation.
2. Roadmap to Cloud	Identify key pain points for the business to solve, then define and align people, processes and technologies to address them based on qualitative and quantitative metrics.
3. Determine Cloud Costs	Review business cases and business requirements documents and ensure that all costs are taken into account so the CCoE can accurately compare costs against expected value delivered for each major solution/service implemented as well as for the cloud service platform. Also do an apples-to-apples comparison of cloud projects to determine the best candidates for a cloud pilot.
4. Calibrate Cloud Vision to reality	Assign portfolio leaders and workstreams to align executive vision (the desired state) to reality (the current state). Goals include identifying the gaps, integrating to workstreams and brownfield solutions, and planning iteratively from vision to reality based on costs and business benefits.
5. Execute and Evaluate	Execute and evaluate portfolio plans approved by the CCoE. Evaluate success early and often based on the specified key performance indicators to feed into continuous improvement. Discontinue projects that don't meet the desired-state outcomes.

SETTING THE STAGE

"Successful digital enterprises are adopting technology roles that are more strategic. For example, the *storage guy* is becoming the *data strategist*. But it's not just a change in title. To succeed in these new roles, technologists must reinvent themselves. They must become trusted advisors for technology, partnering with their peers, providing their expertise wherever it's needed in the business."

—Cindy Schumacher
Chief Information Officer

This book assumes that senior IT leaders have already set the stage for moving to hybrid cloud as part of the enterprise digital transformation. Even if your enterprise hasn't yet addressed every one of the assumptions listed below, you can still launch your hybrid cloud and digital transformation initiatives. However, the more you have addressed these assumptions, the greater your level of success and the faster you will progress in accomplishing the transformation.

Key Assumptions

1. **Business and IT strategies are aligned and intertwined**. Aligning strategies across IT and the business has been a topic of discussion for many years. However, more than alignment, business strategy and IT strategy must be intertwined. One CIO interviewed for the book stated it best: "The business strategy is the technology strategy because we are all working for the same company."
2. **A sponsor from senior management is visible**. As with any major initiative, obtaining the attention, commitment and visible support of senior management is a critical success factor. Direction should come from the top down, not from the middle or bottom up. This book assumes that senior management has already put in place the necessary direction and has allocated the resources to support digital transformation.
3. **Senior management has clearly articulated the company's business strategy and primary objectives and recognizes that business and IT objectives must be closely aligned.** This requires collaboration at the senior executive level. The CIO, who owns the resources supporting the cloud platform, infrastructure and applications must work closely with the executives who own the core business functions such as website, sales and marketing, human resources (HR) workflows and so forth. Their teamwork and cooperation help ensure that the needs of the business are met.

Senior executives must buy into the need for a radical shift in resources to achieve the agility, velocity and high quality demanded by the transformation. This includes the allocation of funds to support transformation efforts such as creating the CCoE and conducting pilots.

Taking a Service Portfolio View

IT's goal in the digital transformation is to build the optimum cloud service platform that delivers the optimum mix of solutions and services to meet the needs of the business. That requires a change in mindset on the part of IT professionals. Instead of a narrow focus on specific products, solutions and projects, IT must broaden its view and focus on the enterprise service portfolio. A portfolio view promotes the sharing of microservices and capabilities, such as user portal, eSignature and search, across multiple solutions and services. Sharing improves development efficiency and increases IT agility in meeting the needs of the business.

Instead of a narrow focus on specific products, solutions and projects, IT must broaden its view and focus on the enterprise service portfolio.

Casting the Roles

You need the right functions in place to support not only the transformation, but also the ongoing business as the transformation progresses. This involves the creation of new functions to augment the capabilities of existing functions. (Although the functions are depicted as individual job roles, they may actually be performed by more than one individual to provide the necessary expertise and experience.) New functions include:

- Chief digital officer – The CDO's mission is to create trust and remove traditional barriers separating technology and business. A primary objective is to balance digital transformation initiatives with keeping the business running and with balancing costs, the portfolio and risks. The CDO function requires both technical and business acumen.
- Cloud service delivery manager – This role owns the cloud service platform from inception to end of life and is responsible for delivering the portfolio of microservices and capabilities that are built into the cloud service platform for consumption by solutions and services that run on top of the platform. The role requires an understanding of business, development and operations. Technology product managers, infrastructure architects and developers with an operations mentality possess the right skill sets to perform this function.
- Cloud security strategist – This role sets policies and priorities for security remediation, trains developers and oversees security initiatives to ensure that security is automated, regulated and preventative. The role requires a background in cyber forensics.
- Cloud compliance strategist – This function is responsible for setting compliance rules and guidelines and training development teams to design required compliance standards into their solutions and services.
- Data strategist – The primary responsibility is to oversee the enterprise data strategy and ensure that the requirements for data analytics, security and compliance are enforced at the data layer. This function requires the skills of a traditional database architect combined with extensive business acumen.

Beginning Your Journey

Now that you have set the stage, you're ready to embark on the journey to hybrid cloud. As you travel on this journey, you may experience roadblocks, delays and detours. But when you successfully reach your destination, you will have elevated the role of IT to a more visible and valued participant in the business.

Figure 8. **Steps to transformation**

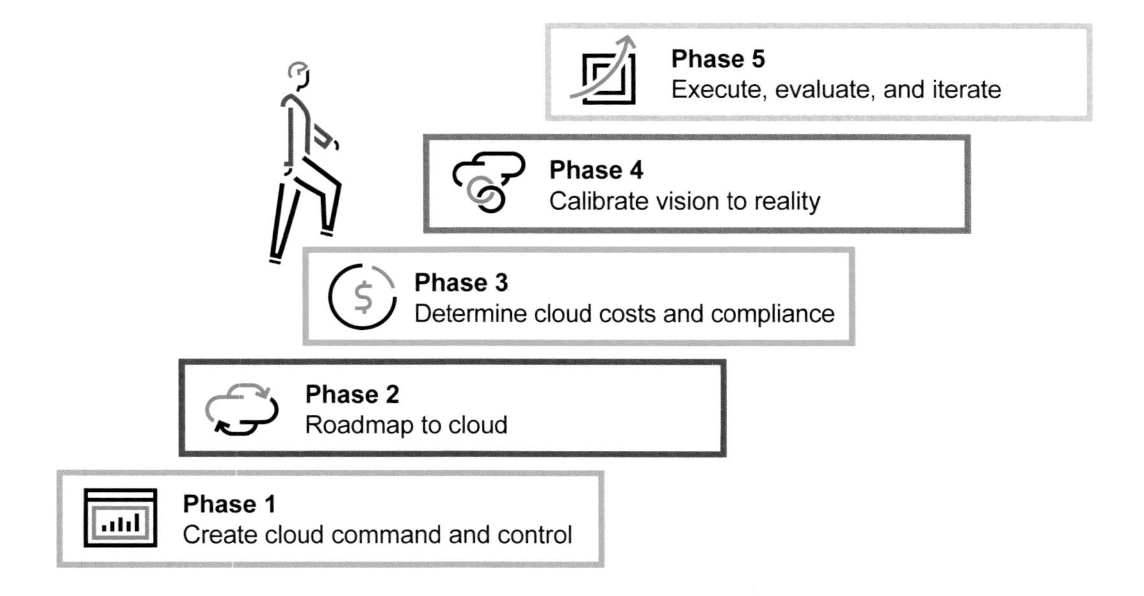

PHASE 1

CREATE CLOUD COMMAND AND CONTROL

"As an early adopter (entirely cloud based for all primary and secondary computing and storage since 2012), leadership discovered the huge communication and cultural disconnect between our business focus (how we make money), and our technology providers (how they make money). We created a fluid communication and decision making process, supported with good measurement and management tools, enabling our internal IT team to let go."

—**Gary Acromite**
Chief Information and Logistics Officer

Phase 1 Objectives

Create an executive-level governance team—a cloud center of excellence—of cross-functional business leaders to define the transformation strategy and create rules and policies to facilitate transformation.

Issues and Clues

Statements that people in the trenches make about their experiences provide insight into the issues that are most likely to block the path to a successful cloud implementation. The following table lists issues the CCoE addresses in Phase 1 and provides actual quotes from IT leaders interviewed for this book.

Issues	Clues
Lack of Alignment	"Our CIO's objectives did not align with those of the COO or CMO. The result is we missed deadlines and revenues fell short of targets." —EVP Business "The technology team was busy cutting key services without consulting business leaders. We had to work around them for our employees to remain productive." —EVP Sales "We implemented a file-sharing application within legal and now the CIO is telling us we have to turn it off because of security risks. Legal owns compliance. IT is just being difficult." —Director of Technology, Legal Department
Lack of Leadership Support	"I own DevOps but I have no visibility into business leadership or the users. I tried inviting them to get together to discuss how they want the system to work, but I am having no luck at getting them to explain what's wrong." —Director of DevOps "The CIO says yes to everything. So we have three public cloud providers and each one has a different interface and process. The business is selecting solutions, but IT doesn't have budget to integrate or support them." —DevOps Engineer

Issues	Clues
Lack of Collaboration	"IT worked with the business to implement a really cool solution mandated from the CMO and CIO. It was shut down after six months of development by the auditors due to security and compliance risks." —Service Delivery Manager "The IT team inadvertently released our price list three weeks early. This blew up our numbers for the quarter because customers decided to wait for lower pricing." —Call Center User
Lack of Visibility	"Users are implementing new solutions that keep hitting us like a log in a game of Frogger. We simply can't keep up with all the new applications." —Engineer "What a joke. The CIO gave the award for the most improved application to an application that the users won't even use. In fact, the users built their own application." —IT Director
Over-engineered Processes	"IT had 35 layers of approvals to integrate and release a service into production. We're talking about going through nearly 350 reviewers before anything gets pushed into production. Both IT and the business are frustrated." —Enterprise Architect
Slow response from IT	"The business wanted updates daily, but the best we could achieve was quarterly because our backend processes were too cumbersome." —Release Manager "IT has been working to update the HR system for six months so we can apply our new policies and give some key people promotions or raises. We're starting to lose good employees, especially those millennials that we really need to retain." —Director Business

Step 1: Create the Cloud Center of Excellence

The CCoE is a team of internal experts whose collective experience and expertise encompass business directives (company, regulatory, security) and an understanding of the available cloud technologies that can support those directives. The team comprises people from different areas across the enterprise, including:

- Key executives from the major business units
- Audit, compliance, security and legal personnel who understand the legislative/regulatory, security and industry requirements relevant to the business of the enterprise
- Technology leaders as well as people tasked with doing the work; One approach for finding the right technologists is to ask the members of the technology leadership teams—development, architecture and operations—to appoint a star player from their respective areas. This combination of leadership and technical expertise helps balance executive oversight and employee input, thereby increasing subsequent adoption.

Every IT leader interviewed for this book was adamant that removing the "I" from IT is a critical success factor. All emphasized the importance of changing the IT mindset from us-versus-them to a focus on company initiatives. The ultimate goal should be to create a spirit of collaboration that facilitates the achievement of well-understood business objectives. Another common success factor is an intense focus on automation, end users (business people) and the absolute necessity for transformation to ensure long-term business success.

Organizations should consider using DevOps as a vehicle for collaboration among operations people, development people and their business counterparts. However, several CIOs interviewed reported they didn't have—and didn't need—a DevOps strategy because their current digital transformation strategies are achieving economies of scale, agility and velocity beyond leadership expectations without DevOps.

CCoE Definition and Purpose

The CCoE consists of a governance and a Scrum component. The CCoE has two purposes:

1. To move high-performing employees to a planning and design role for process automation
2. To create an execution arm to automate the processes determined to be suitable for automation

Down the road, additional competency centers in specialized areas will plug into the overall structure. Specialized areas might include data and analytics, security, governance and compliance, client, IT service management and finance.

Characteristics to Look for

When putting together the CCoE, senior leaders should look for people with specific characteristics. Many interviewees noted that the success or failure of digital transformation initiatives was determined more by the attitude and actions of team members than by technology or process. Effective CCoEs include people who match the following profiles/characteristics:

Characteristic	Description
Proactive innovators	These team members are similar to serial entrepreneurs in that they are the first to recognize patterns, new technologies or changing user expectations. They have a proven track record of making changes proactively, before negative impacts occur.
Early adopters	These team members actively seek new input and solutions to problems. They often experiment with new technologies or ideas in their spare time and are passionate about trying new things. They actively embrace change and thrive on bringing clarity to chaos.
Leaders	These team members have a natural ability to inspire and guide people, teams or organizations. With respect to digital transformation, people look to a leader to drive the vision for new technologies, processes and best practices.
Multimodal communicators	These team members speak confidently in front of business, technology, finance and operations groups. They often serve as translators across teams to ensure inclusion, determine best fit, negotiate requirements and build bridges.
Collaborators	These team members have a track record of building bridges and gaining trust with diverse groups. They are collaborative in style and approach. Their primary focus is building solutions and solving problems instead of chasing political aspirations.

Ensuring Broad Representation

An effective CCoE includes sample representatives from the stakeholder groups listed in Figure 9 and possibly other groups depending on the industry and other factors. The members are the best and brightest in their departments. Think of them as the ambassadors of their respective functional areas. They aren't necessarily the senior leaders, but they are trusted advisors to the senior leaders.

Figure 9. **Summary of functional areas represented on the CCoE team**

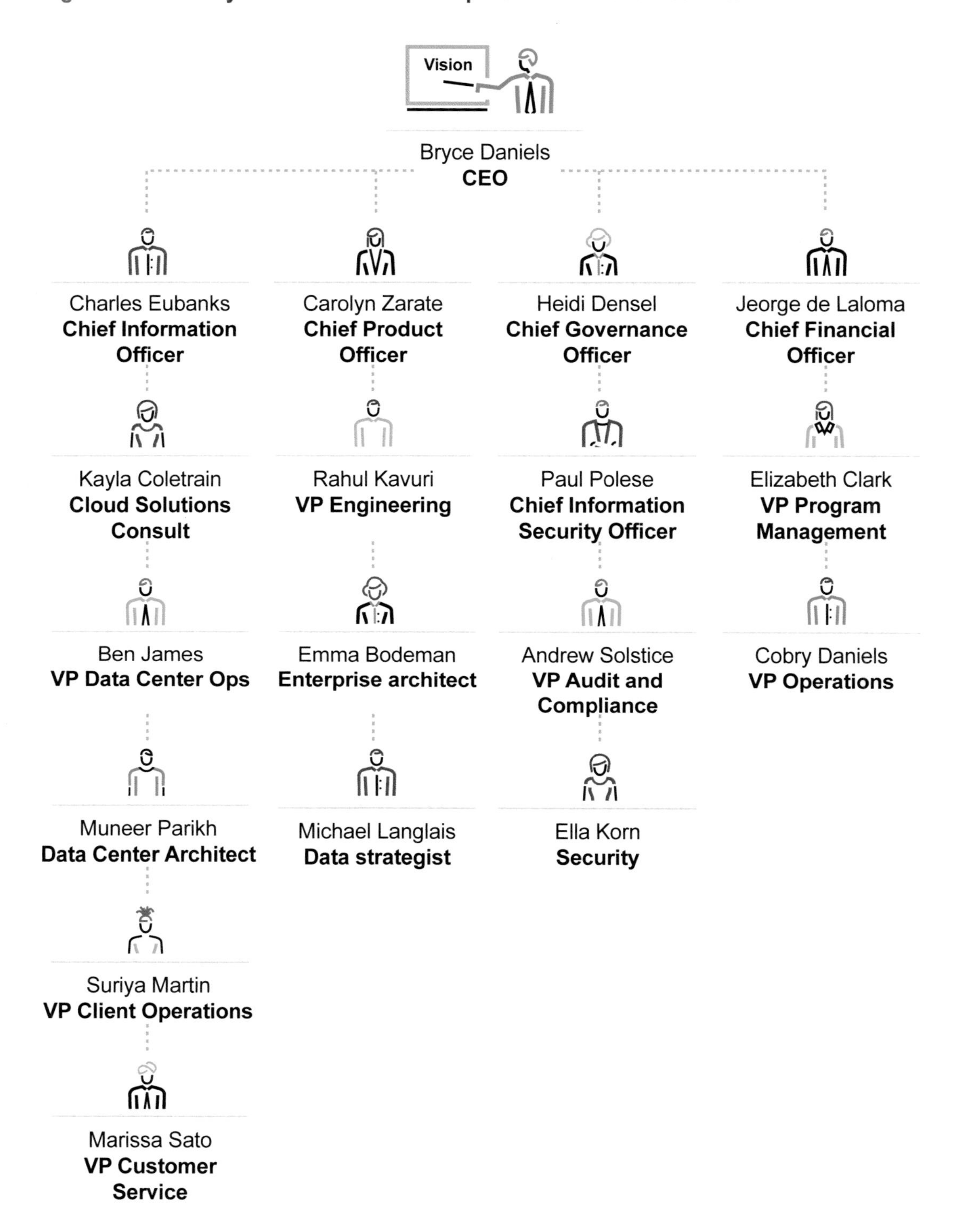

The Role of the CCoE

The role of the CCoE includes the following:

- Identify creative and constructive ways to balance innovation with legacy processes, systems and people
- Create the foundation and rules to automate the many backend processes needed to build a cloud service platform that new solutions and services can plug into
- Identify which processes to keep, automate, eliminate or offload as part of the transformation effort
- Create high-level policies spelling out which workloads belong in private cloud and which belong in public cloud
- Establish overall design principles based on the service profile classification (low, medium or high)
- Break down silos by establishing rules of engagement across all functional areas

CCoEs vary across enterprises depending on such factors as industry, market, consumer expectations and applicable regulations and industry standards. For example, a company that provides a free music service would likely have fewer competency centers within the CCoE than a global financial institution. A large multinational manufacturer or a midsized software company might adopt a multitier structure in which a core CCoE drives people-process-technology decisions and change at the global level. That top-level team then creates second-level CCoEs in each country or region to provide country-specific governance and market requirements and to serve as execution teams that create microservices for their respective regions or countries.

The CCoE Scrum Team

Vision without execution is delusion. Too often, tactical teams are put in place to execute a change project but the team members are still burdened with their day jobs. As a result, team members usually don't have the time to take on both responsibilities, so they don't do either job well. As a result, projects are stalled or stop short of achieving the expected vision and goals.

Scrum is an iterative and incremental agile software development framework for managing product development.

To eliminate this problem, the CCoE should include a dedicated functional Scrum team to create microservices, APIs and automations that form the foundation for automated solutions and services. The team includes both developers and operations engineers and serves as the liaison between traditional operations and development teams. The Scrum team provides a "middleware layer" between the cloud design teams and traditional solutions to eliminate inertia, accelerate time to value and maintain compliance to key directives related to security and regulatory requirements and corporate policy. Ideally, the Scrum team reports to the cloud service delivery manager who is ultimately responsible for driving the design and building of the cloud service platform.

Step 2: Establish Rules of Engagement

For years, IT has talked about the people-process-technology equation. However, too many projects focus on process and technology and ignore the people component. That's why so many cloud projects and digital transformation initiatives fall short of the mark.

The CCoE must create an environment in which all participants are treated as equals, people are held accountable and projects move ahead. Rules of engagement create that environment by specifying how people are to work together. Well-thought-out rules ensure that transformation initiatives don't turn into unruly meetings with scores of attendees vying for control. Figure 10 lists some important and proven rules of engagement.

Figure 10. **Rules of engagement that promote collaboration and drive success**

Rule	Description
Respect for differences	Everyone understands that each team member brings value to the conversation. Team members focus and collaborate on solutions and not on personal agendas or problems.
Lean team	Limit participation to one representative from each affected area to keep the team from becoming unwieldy.
Equal voice	Everyone has an equal voice and vote in discussions and the cloud service delivery manager acts as the tie breaker to make the final call for decisions that do not reach consensus.
Escalation pathway	The goal should be to resolve disputes within the team. However, an escalation path to the executive level must be clear in case the team can't resolve an issue or needs executive approval to proceed.
Axe arsonists	Anyone who is too political or who constantly creates unnecessary churn that hampers progress can be replaced by someone who will do the job more effectively.
Responsibility and account-ability	Meetings are managed for maximum efficiency. An agenda drives each meeting, individuals report on progress toward goals and action items are assigned. Individuals who consistently show up for meetings unprepared will be replaced.
Ownership	Ownership for requirements and/or competency centers should belong to the person who has the authority to allocate budget or resources for the solution. For example, the owner of a security competency center would be the person appointed by the chief information security officer.
Rule of three	The first step is to ask individuals to select a preference from a list of alternatives. If they don't respond to the first step, the second step is to inform them of the group's preferred choice. If they don't respond to the second step, they are then informed of the final decision.

Rule	Description
No hidden agendas	Hidden agendas are checked at the door. Every participant needs to focus on balancing the portfolio and resources against corporate goals, not personal or departmental goals.
Insulation from inertia	Members must be present not just physically but also mentally during meetings. No email, texting or side conversations except during breaks. Cloud initiatives are critical, so senior leaders need to prevent interruptions to CCoE meetings.

Step 3: Address Security and Compliance by Design

Even projects with close collaboration across business and technology groups can fail if key elements are not included early in the planning process. As the following examples show, failing to incorporate security requirements from the beginning often results in spending millions more than expected and introducing serious business risks.

- In February 2015, Anthem was hacked and nearly 80 million records were stolen. The aftermath of the breach has led to a class action lawsuit.[11]
- In March 2015, a federal judge approved a $10 million settlement of a class action lawsuit against Target based on a 2013 data breach that affected 40 million customers.[12]
- In February 2016, a U.S. hospital's critical patient care systems were taken down by a hacker and weren't restored until the hospital paid a ransom.[13]

To prevent breaches such as these, ensure that security is front and center in discussions and an integral part of core planning and architecture design. Incorporating security and compliance throughout the process eliminates concerns about the impact of breaches later, which includes the cost of responding to the breach as well as brand damage and potential financial penalties.

Top performers in digital transformation proactively create microservices that continually check applications and configurations for vulnerability to breaches. The ransomware attack on the hospital cited above could have been avoided if more automated checks and balances had been built into access and approval controls from the start.

---

11. "One year later, controversy still surrounds Anthem data breach," Leslie Small, *Fierce Health Payer*, February 2016. http://www.fiercehealthpayer.com/story/one-year-later-controversy-still-surrounds-anthem-data-breach/2016-02-23
12. "Target will pay hack victims $10 million," CNN Money, Charles Riley and Jose Pagliery, March 19, 2015. http://money.cnn.com/2015/03/19/technology/security/target-data-hack-settlement/index.html.
13. "Hollywood hospital pays $17,000 in bitcoin to hackers; FBI investigating," Richard Winton, *Los Angeles Times*, February 18, 2016 http://www.latimes.com/business/technology/la-me-ln-hollywood-hospital-bitcoin-20160217-story.html

The figure below is a good illustration of a security risk assessment performed by the security strategist to determine the high level policies required for storing protected data across various clouds and devices.

Figure 11. Security policies

Cut costs, compliance, and project risks by conducting a security assessment and creating policies based on data location, access and risks to the company.

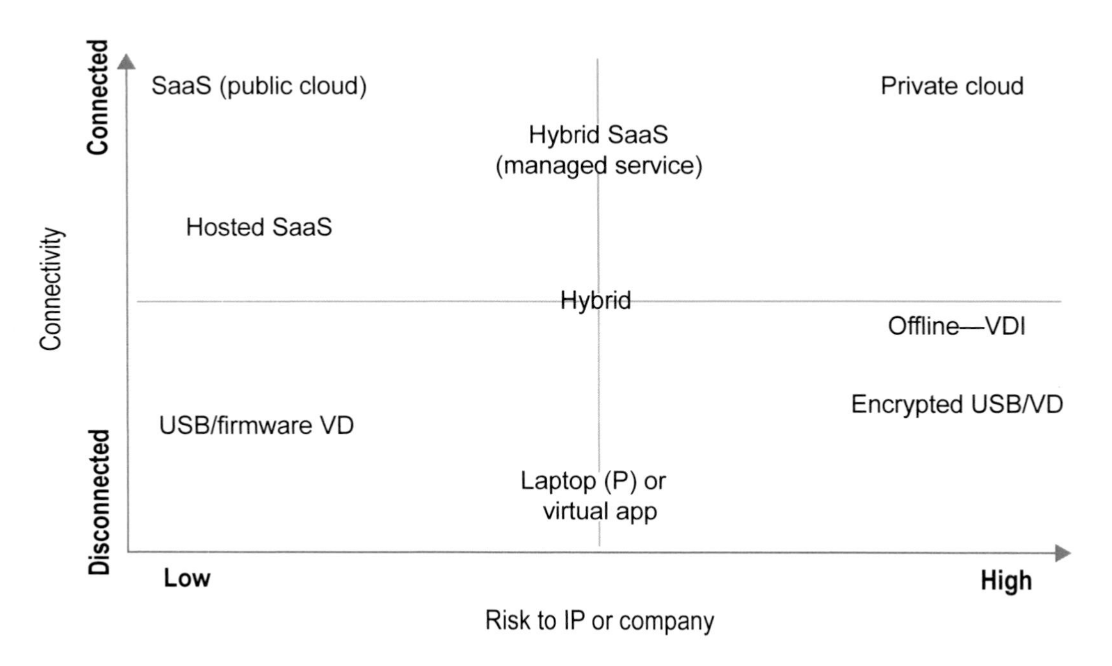

With this in mind, the CCoE must direct technology teams to incorporate programmatic layers of controls for critical changes such as automatically encrypting a database. The hacker that took down the hospital patient care system encrypted the database used by that system, thereby preventing the system from accessing the data it needed. If the hospital's security policy had required at least three levels of signoff for encryption to be applied, the hacker would have been less likely to have been able to encrypt the database because a manual override would have been required.

When addressing security, keep in mind that business people focus on getting their jobs done efficiently and productively. Security policies that are overly cumbersome hamper productivity and may tempt people to look for workarounds:

- If you don't provide an easy way to securely transfer large documents, business people will inevitably turn to popular file transfer services.
- If you don't provide secure internal chat, they'll use WhatsApp, Facebook Messenger or SMS.
- If you restrict email size to 50 megabytes, they'll use gmail.
- If it takes six months to deploy a virtual machine (VM), they will turn to unmanaged public cloud solutions and pay with corporate cards.

The key is to find the right balance between security and usability. A process that constantly solicits employee input to gain insight into service needs and ensures that those needs are met helps bring shadow IT activities under control.

Step 4: Create Guardrails

CCoE members must establish a structure that provides *guardrails* to ensure that projects adhere to the rules and policies of the governance framework. That structure includes competency centers tasked with developing requirements for specific functional areas. Examples of competency centers include:

- **Customer-facing development** – Provides requirements for customer-facing services, a role that has traditionally been held by product management and enterprise architects in the business.
- **Cloud service delivery team** – Comprises operational team members who manage current virtual desktop infrastructure (VDI) and cloud solutions.

CCoE members must establish a structure that provides guardrails to ensure that projects adhere to the rules and policies of the governance framework.

- **Security and compliance** – Provides minimum requirements for automation rules that address both security and audit needs, thereby ensuring security and compliance are designed into cloud solutions.
- **IT service management** – Provides process maps and descriptions of the current state of ITSM implementation and tools. This competency center provides, for example, details of the current change process along with what needs to remain intact, what needs to be automated and what systems need to be touched.
- **Finance and operations** – Provides guidance on profit and loss cost models, depreciation cycles for assets, financial implications of selections and any requirements for financial or operational systems.

Competency centers play an important role but they should be limited in number.

Step 5: Create a Cloud Positioning System

Before the CCoE digs into assessing the current state of cloud within the enterprise, members must establish a standard for cloud-based solutions and workloads. Part of assessing the current state is understanding where workloads belong. Creating the baseline before digging into the details enables the CCoE to properly frame and prioritize next steps.

A cloud positioning system is a high-level chart that guides decisions regarding where certain workloads should live. The chart serves two primary purposes. First, it enables the CCoE to quickly identify candidates for the cloud. Second, it is used to programmatically determine the options presented to users in the self-service portal for requesting as-a-service resources.

In creating the cloud positioning system, it's a good idea to use hypothetical solutions and services based on real-world workloads instead of company-based workloads that might be politically charged. A workload is not necessarily an entire solution or service.

It could be one of the multiple components or microservices that comprise a solution or service. The strategy helps determine where each workload is to be placed.

As Figure 12 shows, you need to weigh three characteristics—time to value, security/ compliance and cost—when deciding where to place the workload.

Figure 12. **Cloud calibration chart**

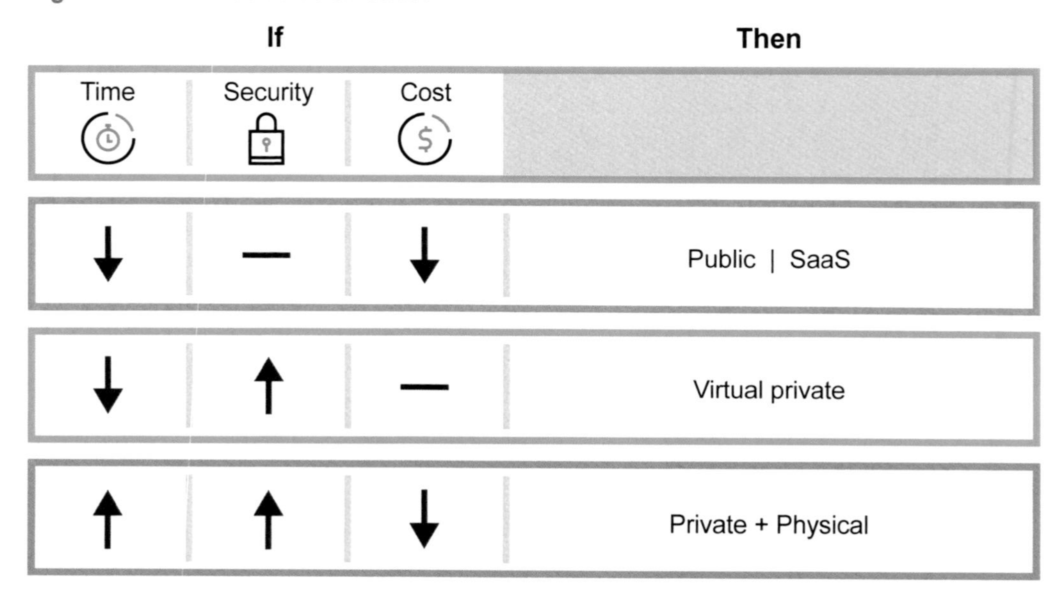

The cloud positioning chart helps determine where a workload should reside.

1. **Time to value** – Some business leaders may call this *velocity* or *agility*. The faster you can release a solution the sooner you start seeing a return and the greater that return is likely to be. When fast time to value is essential for seizing a market opportunity, then public cloud or SaaS might be the best choice if current internal processes take too long.
2. **Security and/or compliance** – Platform choice is affected by many different security and compliance considerations. Governments around the world mandate that companies protect confidential and sensitive data. In some cases, legal mandates require that data remain within the country of origin. In politically unstable countries, companies may insist that data backups reside in a country with a more stable

political and economic environment so that systems and data can be recovered if political unrest results in data seizure or loss. For this reason, companies often choose to run applications with highly sensitive data internally, either on a traditional IT infrastructure or an internal private cloud.

3. **Cost or financial impact** – This factor relates to costs, savings and budget. Some solutions and services have intangible benefits that cannot be easily measured but are critical such as a new service for which there is insufficient data to predict impact.

You may also want to address "people" issues. Although people issues weren't as prevalent in 2014, they now come up quite a bit more in interviews. Many leaders have reported moving workloads to the cloud or a SaaS solution because of skills or resource retention concerns. Some top performers lump the resources/skill factor with the cost factor while others call it out in an additional column for additional guidance. There is no right or wrong answer here. Just be sure to keep it simple.

Step 6: Assess the Current State

The final step in Phase 1 is to collect information about current enterprise cloud initiatives and solutions so you can develop a clear picture of where the enterprise is with respect to cloud. The information lays the foundation for Phase 2, which focuses on creating a cloud roadmap.

Generally, members should be responsible for collecting details of the cloud initiatives within their respective functional areas. Collecting the data takes time and CCoE members need to know where to find it. Here are a few hints on where to look for data:

- The configuration management database for solution/service mapping data and VM instances
- Expense reports and discovery tools that track URL requests to find shadow IT implementations
- Logging tools and security tools for information on security vulnerabilities and settings. A security assessment may be necessary to identify the biggest errors in programming and what solutions could be there to address them

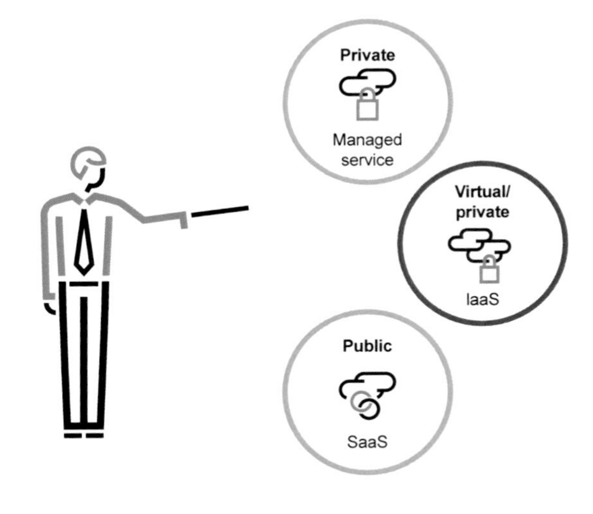

- Logging tools for help in identifying service mapping

Current state descriptions should remain at a high level and include:

- Proposed initiatives involving cloud technology
- In-flight initiatives along with the associated forecasting/business cases describing expected results
- Cloud audit findings for shadow IT
- Current cloud vendors in play for infrastructure as a service (IaaS), platform as a service (PaaS) and SaaS
- Integrations, with hot spots and cold spots
- Current processes for IT service management
- Service level agreements
- Security profiles and policies
- Compliance restrictions and constraints
- Financial modeling templates (needed for Phase 3)
- Current key performance indicators (KPIs) and assessment, if available (for example, cost per minute of downtime for a given service and average downtime for that service)
- Examples of gaps in all the above
- List of immediate services that are relatively low risk for onboarding
- Current process maps for business processes, beginning with currently funded business and/or IT projects

Hot spots occur when resources are overutilized causing bottlenecks due to a large queue of requests. *Cold spots* occur when resources are underutilized due to a failure to identify or accurately predict velocity.

Best Practices Top Performers Use to Assess the Current State of Cloud

1. **Look for brownfield not greenfield** – Unless you are a startup, your current state map should depict a brownfield environment with a number of legacy, on-premise and third-party cloud solutions.

2. **Think outside the box** – Dig deep and uncover every occurrence of shadow IT. Then determine the best way for the company to leverage what it has to move forward. Pull data from expense reports in addition to using traditional discovery tools. The CEO of a major technology manufacturer decreed that anyone who fails to declare a shadow IT incident is subject to termination.
3. **Stay at a high level** – Diving into the minutiae, legacy baggage and political agendas can sidetrack the discussion.
4. **Paint an honest picture** – Get an accurate picture of the current state so you can address issues instead of covering them up or perpetuating them. For example, an Intel audit uncovered 1,800 shadow IT applications. IT brought shadow IT into the light and leveraged it by creating a process that enables business people to order what they need from third parties while still ensuring appropriate tracking.[14]
5. **Focus on business capabilities, not technology** – Consider company objectives and the solutions/services that map to those objectives *before* looking at the technologies and solutions that map to those services.

Figure 13. **Summary of Phase 1—Create cloud command and control**

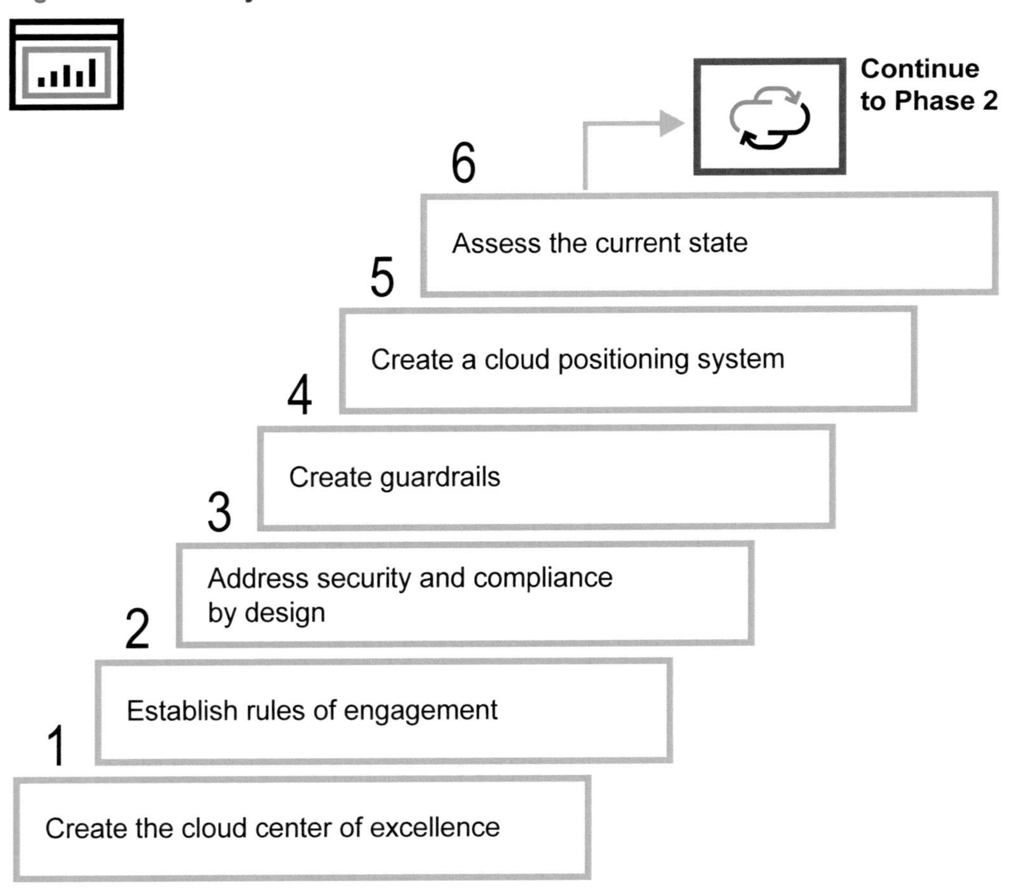

14. Bolton, Clint; "Intel CIO Finds Business Value in Shadow IT." June 8, 2015, *Wall Street Journal.* http://blogs.wsj.com/cio/2015/06/08/intel-cio-finds-business-value-in-shadow-it/

Words of Wisdom from the Trenches

Aaron Amendolia
Vice President, IT Service Delivery,
National Football League

The Secrets of Successful Transformation

Transformation Shift

Technologists need to transition from the traditional role of shipping boxes and systems to a more strategic role. If they cling to legacy solutions or expertise in silos, they will not survive the digital transformation. High-performing employees evolve into strategic roles and rebrand their value to the business. They don't think of themselves as storage guys but as data strategists.

Cloud service delivery leaders focus on helping business leaders succeed with hybrid cloud solutions. They become trusted technology advisors who break down operational silos. One minute they could be working with legal for litigation or product management for consumer-facing apps, and the next minute they are pushing the cloud vendor to comply with service level commitments.

The transformation changes the type of people who succeed. People who don't embrace the transformation find themselves looking for new careers or working for the cloud providers.

Technology to Governance

The effort should include building out risk and compliance solutions for technology. It has to include people fluent in all aspects of the business, not just technology. Strategists should shift attention away from the *technology* of the systems and focus on the *governance* of the systems. The governance team (the CCoE) should concentrate on having the right agreements and solutions to ensure compliance and security. Success in the cloud depends on having vendor agreements that are fully vetted and in place. The team expands to include many areas, from technology and business to legal, compliance, finance and risk analysis. All members should become well versed in data governance, that is, what happens to the data and who has access to it.

Trusted Advisors

Trusted advisors come from within. They are people who have a vested interest in the success of the company and its customers. Trust is built through actively listening to customer challenges and providing solutions that address those challenges. Saying *no* or quoting policy will only encourage people to look for alternative solutions. Embracing suggestions enables leadership to keep up with the pace of change and make better decisions *with* the business not *for* the business.

PHASE 2

ROADMAP TO CLOUD

"The business requirements document proved to be an invaluable tool for working with stakeholders. The different areas covered in this document helped galvanize disparate teams, forcing them to develop actionable goals. Completing this important step during the ideation phase allowed for smoother and quicker execution and less churn, as it was repeatedly cited."

—Tom Klemzak
Management Consultant

Phase 2 Objectives

Define business requirements for the cloud service platform and understand requirements for solutions/services that will run on that platform. Create an initial list of pilot candidates for the cloud service platform.

Issues and Clues

The following table lists issues the CCoE addresses in Phase 2 and provides actual quotes from IT leaders interviewed for this book.

Issues	Clues
Unbalanced Portfolio	"When developers follow the cool shiny objects and there is no clear business benefit, projects fall apart or get killed. Leadership needs to emphasize the importance of balancing new technology against our accumulating technical debt." –Enterprise Architect "Hard set rules with multiple silos and repeated technology efforts are resulting in disconnected data, redundant teams and lost opportunity." –Vice President, IT Service Delivery
Not Trusting Internal Experts	"All software has problems. Unfortunately, buyers often don't believe us when we tell them that—until they need us to fix the software or integrate it with other systems." –Cloud Service Delivery Manager "SaaS and other cloud providers are overselling capabilities to my peers in the business. After the contracts go away, the technology team is stuck trying to figure out how to make up for areas where the system falls short. But we aren't given the resources or budget to fix the problems." –CIO
Poor Planning	"Consultants come in with best practices and implementation advice as the trusted advisor without understanding the unique needs of our business. Rarely do the solutions work out of the gate and more often than not, after the consultants are gone, we are tasked with fixing the issues." –Cloud Service Delivery Manager "Every time we turn around, we're handed another application to tie in, but management isn't giving us any additional budget to do it." –Director Data Integrations

Not Including all Stakeholders	"The project started with nine applications and just a few integrations. Now we're struggling to keep up with requests to integrate 30 more applications that absolutely must have data from or share data with the new solution." –DevOps Engineer "Tempers are flaring. Our VP of infrastructure operations threw the company Christmas tree at our CIO because the vendor selections were made without including his team." –Director of Engineering
Lack of Standards	"Our biggest challenge with hybrid strategies revolves around security and understanding how it relates to data governance. Traceability, accountability and accessibility are essential, but the challenge of building them in is becoming more and more daunting because we are not called in until after production rollout." –Security Engineer "We have three cloud providers with three different sets of interfaces and APIs to integrate into. Instead of calling it bimodal IT, analysts should call it bipolar IT. It's hard to keep up with the pace of change across the various clouds, tools and requirements." –DevOps Engineer

Step 1: Prepare to Take the Optimum Route

Imagine a restaurant serving up a whale and expecting people to swallow it whole. But cut it up into bite-sized pieces, season it and present it in an appealing way, and you have sushi that's easy to consume.

CCoEs are sometimes tempted to take a "whole whale" approach that encompasses every proposed cloud solution, horizontal service or dependent application. That's because the enterprise already has cloud-based solutions in place as well as ongoing cloud projects. However, it's likely that previous cloud projects have had limited oversight. Consequently, a lot of them may be stalled and in need of attention.

Instead of attempting to address all the disparate cloud-based solutions at once—the equivalent of trying to swallow a whale whole—keep sushi in mind when defining the cloud roadmap. Start with a small number of manageable projects. Work with business leaders to review the top 10 solutions/services that align closely with the company's strategic goals. Meet with each solution/service owner to understand the business requirements for each solution or service.

In developing the roadmap, you'll find that many people—including executives and leaders—want to weigh in given the importance and their own interpretation of cloud. To avoid mayhem, do your homework by researching what the company has already done, what successes have been achieved, what cloud solutions people are championing and what their pain points are before you solicit their feedback.

Focus on the Solution, not the Technology

People sometimes become enamored with a technology that was successfully implemented in another company, business unit or department. They may chase after that technology, trying to apply it in other situations without taking the time to develop a solid understanding of the business requirements. For example, during a process mapping session at a large manufacturer, the facilitator decided on the "right" solution before the group had defined the processes and requirements and was able to convince business leaders to conduct a proof of concept (POC) before they were ready. During the POC, it became obvious the solution couldn't address 90 percent of the requirements. Unfortunately, the team wasted a lot of time and effort piloting a solution that was completely inadequate.

Get User Feedback Early on

Often executives and directors take the lead on defining processes and requirements for new solutions/services and forget to involve the people who have the most at stake. These are the people who are doing a job every day. They truly understand the processes in place, what's wrong with them and what capabilities the new solution/service needs to provide.

At one company, executives spent a morning mapping the current process for customer relationship management and ended up with a detailed process map on a whiteboard. During a break, a senior vice president on the business side of the house brought in people who would be the primary users of the proposed solution/service and asked them to review and correct the process map using sticky notes. Very quickly, the whiteboard was covered with colorful squares. The input from these people resulted in a significant changes, not only in the process but also in the technology, solution and rollout strategy.

Step 2: Review Business Requirements Across Solutions and Services

The CCoE must develop a thorough understanding of the business requirements for the cloud service platform and the solutions and services that might be candidates for running on that platform. To do that, the CCoE relies on the business requirements documents (BRDs) developed by the solution/service owners.

A BRD is a blueprint that documents the basic structure and requirements of a solution or service, which then enables the owner to get management approvals,

A BRD is a blueprint that documents the basic structure and requirements of a solution or service estimate costs and guide the development. Although companies typically have BRDs for customer-facing services, this critical document typically does not exist for internal IT projects. The CCoE needs to ensure that the cloud service platform and other internal IT solutions and services have BRDs that map to the relevant business cases describing the costs and benefits associated with the platform, solution or service.

> "Just as you need a blueprint to build a house, you need a blueprint to guide the introduction of new solutions and services, regardless of the development style or method your company uses. It's the only way to ensure a solid foundation and structure across stakeholders."
>
> — Pat Spica
> IT Director

Typically, the cloud service delivery manager writes the cloud service platform BRD and collects solution/service BRDs from their respective owners. If a BRD hasn't been written, the cloud service delivery manager should work with the solution/service owner to ensure one is written. It typically takes no more than a week to collect details from the BRDs.

The BRD template at www.ispeakcloud.com has been used for more than two decades and it drives success in many industries and many company sizes ranging from startups to enterprise IT environments.

How the CCoE Uses BRDs

CCoE members leverage the BRDs to begin understanding the business requirements for various solutions and services so they can identify the best candidates for the piloting of solutions/services on the cloud service platform. BRDs help the CCoE create alignment across critical elements from stakeholders, including:

- **What** – A definition of the business requirements that the cloud service platform must accommodate for multiple solutions and services
- **When** – The timelines and schedules within which a solution or service must be developed and rolled into production
- **Why** – Insight into why the enterprise is investing in the ssolution/service and into the primary objectives and KPIs that will be tracked to demonstrate success
- **Who** – What resources and skills are required from the business and IT to implement the solution
- **How** – People, process and technology flows that must be documented and linked from the BRD or included within the BRD across the value stream

Identify Baseline Services and Microservices

The CCoE leverages the BRDs to look across all solutions and services and develop a set of common microservices that those solutions and services must have in place to function properly. These include non-Agile requirements related to reliability, availability, serviceability, supportability, scalability and security. The cloud service delivery manager is responsible for ensuring that these requirements are addressed in the cloud service platform so that solutions and services that run on the platform can take advantage of them:

- **Reliability** refers to expectations in terms of service load and requirements to maintain the system for business continuity and disaster recovery. The service level categories might include *gold, silver* and *bronze*.

- **Availability** defines the required time for the service to be up and running. Availability varies with the service, number of users and times of use (daily, monthly, quarterly, yearly). It could be as simple as *the website needs to have a 99.9% up time or the service needs to be available only at end of each month.*

- **Serviceability** sets expectations for the frequency and requirements for servicing the application. It also defines what hours/times are acceptable to support business requirements. It answers such questions as: Are there blackout periods when the system cannot be touched? Are daily, weekly or monthly updates required for pricing or other business-related changes? How many patch updates are anticipated and what is the frequency?

- **Supportability** is the desired support model at go-live. Will you have tier 1 (help desk), tier 2 (advanced troubleshooting) and tier 3 (development level) support? Or is the product mature enough for a partial tier 1 call and resolution by development as needed? Can support be automated with a virtual assistant for tier 1 issues or will support agents be necessary?

- **Scalability** refers to the number of users, forecasted data load and other elements that can affect networking, data, storage and system performance. This information helps in planning for load testing, network usage and other elements based on similar solutions and services. Scalability is often an overlooked but critical aspect of any cloud model.

- **Security and compliance** includes legal mandates to protect data, personal information acts that vary from one country to another and rules around the location of data and backup systems for business continuity. Identifying security and compliance requirements in advance enables you to architect them into solutions, services and the cloud service platform. Studies show that once you're in production, it can cost 100 times more to address a missing requirement than it would have cost if it had been part of the original design.

In Phase 1, you created competency centers to address some of these requirements and the service portfolio process takes into account requirements, policies and rules. If competency centers do not already exist, the CCoE should request that the leaders of core horizontal areas such as security/compliance and data assign service portfolio leaders to define the rules and categories needed for the cloud service platform BRD.

Connect Solutions/Services to Corporate Objectives

The CCoE develops a common understanding across all stakeholders on what is a minimum viable product and how each solution or service ties into high-level company objectives:

- Each high-level capability stems from a company objective (for example. enable electronic signatures).
- Each objective ties to a business case that provides a value metric for the functionality (high, medium, low):
  - Anticipated savings
  - Anticipated revenue
  - Business requirements to achieve those goals:
    - Skills (people)
    - Solution (process)
    - Scope (technology, timelines, users)
    - Success (metrics)
- Each capability ties back to technology solutions from the technology team based on suggested areas of enablement.
- If some capabilities require major architectural considerations due to security, compliance or costs, the architecture team may require an additional level of detail to tie it back to the governance framework and cloud service platform.

> Enterprises with hundreds of solutions and services in their service catalogs should create BRDs only for initiatives that align with current business objectives.

Enterprises with hundreds of solutions and services in their service catalogs should create BRDs only for initiatives that align with current business objectives. They may opt to migrate out-of-the-box solutions with minimum changes directly to a cloud infrastructure based on the high-level policies created for workloads during Phase 1. In time, all new solutions or services listed in the service catalog or major upgrades to previous ones should have a BRD.

Uncover Patterns

The review of the BRDs provides insight into whether new cloud solutions are custom built or off the shelf. The BRD reveals patterns across solutions and services so you can identify opportunities for automation, elimination, facilitation and consolidation.

- **Automation** – Provide automation via microservices for key solution/service elements. Begin by automating core operations functions such as initiating a self-healing operation with an orchestration tool in response to a monitoring alert. Other popular automation targets include change request, change order and update processes; update of the configuration management database; and open and close of incident and problem tickets.
- **Elimination** – Identify legacy systems that are being replaced by a new solution/ service. For example, implementing a reporting and analytics microservice that solutions and services can plug into may eliminate the need for one-off reporting tools, multiple dashboards and/or integrations. Elimination helps achieve cross-service savings.
- **Facilitation** – Reduce spend, enable security/compliance and accelerate time to value by providing a standard solution for applications to plug into. One example that came up regularly in interviews is a rules-based open source tool to check solutions and services for adherence to security and compliance requirements. If a solution or service passes the check, an automated approval process kicks in, provided budgetary approvals are in place. If the solution or service fails the check, the tool sends a list of necessary adjustments to the developer and product owner so they can take action to move the solution or service to the next step. (If the adjustments can't be made, at least the owner and developers understand why the solution or service can't move forward.)
- **Consolidation** – Consolidation of tools, providers and solutions across the technology stack delivers benefits to the business and technology groups. For example, the IT organization in a large technology company was faced with satisfying the demands of the business for rapid adoption of cloud technologies. The CCoE found a large volume of workloads on Amazon Web Services (AWS) scattered across the enterprise. The team used that knowledge to negotiate a 30 percent discount with Amazon, which reduced total cloud spend.

Don't get bogged down by hype or nomenclature. Different companies use different terms for the platform and associated blueprints to build out the supporting infrastructure, and new terms continue to appear. Focus on creating a common language, a standard set of requirements and an implementation plan that works for your environment.

With the completion of this step, you have gathered all the information the cloud service delivery manager needs to produce a BRD for the cloud service platform and you have insight into the business requirements for the various solutions/services that you are considering as pilot candidates.

Step 3: Assess Risks and Identify Gaps that Affect Objectives

In this step, you conduct a risk assessment and gap analysis in people, process and technology areas to identify anything that might adversely affect the success of either the solution/service or the cloud service platform.

People

Every interviewee remarked on the importance of the human element in cloud success, and indicated that people represent the biggest source of risk in getting any solution to market. Consider the following factors as you work through the people-process-technology equation.

People represent the biggest source of risk in getting any solution to market.

- **Mix of experience level** – The solutions/services you implement need to accommodate employees of different ages and experience levels, from the least skilled to the most skilled. Bringing millennials into the process enables you to not only address talent retention concerns but also to leverage their insight into how solutions/services should evolve. A good way to do that is to assign millennials to projects that have lower levels of business risk. This enables them to learn the ropes by participating in the launch of a new solution/service, thereby preparing them to handle higher-profile projects that have a greater business impact. In this way, the enterprise can balance innovation so that the solution appeals to both audiences. Some top performers assign mentors to guide up-and-coming talent, an approach that appears to have a high success rate.
- **Skill shortage** – Many CIOs are struggling to find people who understand technology, understand the business and have the ability to effectively communicate plans and architectures. Some CIOs are outsourcing the cloud service platform to a third party instead of trying to build skills internally. Other CIOs are hiring the key resource—the cloud service delivery manager—and then leveraging crowdsourcing or outsourcing to provide specific architectural and technical skills. CIOs in larger enterprises that can afford the higher salaries of multifaceted resources are building out teams to facilitate digital transformation. Some CIOs report increasing their investment in training to help employees with good track records transition to new roles and acquire needed skills.
- **Motivation** – The skill shortage has made retaining and motivating top talent more difficult. Some top performers are experimenting with motivators such as offering charitable donations in the employee's name instead of giving the employee gifts. One successful technique involves turning superusers into champions of change among their peers. This gives them a voice and enables users, technologists and business people to overcome many of the communication challenges.

- **Fear, Uncertainty and Doubt (FUD)** – Two-thirds of interviewees cited the FUD factor as the greatest inhibitor to progress. FUD makes it difficult for employees who have built a career on their expertise, such as the *storage expert* or the *server expert*, to transition from a reactive hero-on-a-white-horse role to a proactive planning role. Top performers are finding creative ways to eliminate FUD. For example, a sports company is revamping job titles and putting people through training to help them transition to the demands of the digital enterprise.

Process

In traditional companies, IT processes have evolved over the years and are codified in standard process frameworks such as The Open Group Architecture Framework, ITIL, International Standards Organization and IT4IT™. People involved in creating these standards understand the value of sharing proven best practices that companies can adopt instead of spending time and money creating processes from scratch.

For younger people who didn't live through the chaos that was prevalent in the wild technology world before process brought order, process frameworks seem cumbersome and redundant. The first inclination for these people is to reject processes that appear to hinder progress.

Companies that have successfully increased agility and velocity while keeping areas such as shadow IT, cloud sprawl and stall in check listened to both sides of the process debate. They then came up with a way to create processes that maintain order and control without hampering productivity. Doing so empowers the enterprise to make the transition to cloud and the transformation to digital. CCoE members who follow these recommendations will achieve that goal:

- **Automate, as long as the processes are as streamlined and fine-tuned as possible** – Interviews with business and technology executives show that many companies have too many layers of process. Automating bad processes creates chaos, so process owners and engineers need to examine and streamline processes before automating. For example, well-established network standards have enabled companies to connect an incredible number of devices together, forming the basis for cloud computing. By applying these standards to other areas such as security, change, configuration, incident and problem management, companies can significantly decrease costs and increase velocity.

> "The ITIL framework was created to be adapted with the ebbs and flows of technology. Although many aspects can be automated as part of the overall cloud service delivery framework, tracking, understanding and analyzing change, incidents, and problems within the technology stack is as critical today as it was 40 years ago."
>
> — Malcolm Fry
> ITIL Luminary/Author

- **Emphasize software-defined change and configuration** – Encourage developers to build core change and configuration management functionality into microservices so human intervention isn't required for every action. For example, when a monitoring tool detects latency or a nonresponsive application, it can trigger an orchestration tool to run a microservice that activates a self-healing process.
- **Automate or adjust change management** – Documenting change is an essential part of maintaining visibility into the environment and controlling drift. Many change management processes were developed prior to the emergence of automation and cloud computing. They need to be enhanced to perform effectively in the cloud. Automate changes that don't require human intervention—for example, adding a button or changing a color—so people can focus on the changes that have a broader impact.
- **Simplify change processes** – Take the opportunity to reduce the complexity of changes that require human intervention. A large software manufacturer did this in a big way and achieved great results. Initially, the company required 35 levels of approval even for cosmetic changes such as the color of a button or a phrase on a website. Senior leaders moved enterprise architects out of IT and into the business to align them more closely with the needs of the business, a reorganization that reduced the number of approval levels. Additionally, they eliminated the requirement to send minor business-related changes through the change advisory board. Minor changes can now be released with quick approvals from the product owners group. Finally, senior leaders created a rules engine that automates many of the manual checks that the change advisory board was doing, which eliminated a number of unnecessary layers.

Figure 14. Embracing Shadow IT Top Performers embrace Shadow IT by working with the business on an automated framework to reduce the impact to velocity and adoption.

Technology

In enterprises with brownfield IT environments, the CCoE must address and attempt to eliminate technical debt—that is, outdated legacy systems and nonstandard technologies—in the roadmap. The roadmap should include plans for running legacy systems and new platforms in parallel while gradually moving more users and solutions/services to the cloud. The following guidelines help in addressing technology issues in your roadmap:

- **Address end of life** – Any new service rollout should include a plan to retire the legacy system that previously provided the service. Get buyoff from stakeholders on the point at which you can completely cut over to the new service and shut down the old one.
- **Eliminate technical debt as part of cloud adoption strategy** – Eliminating technical debt is a key driver to adopting cloud-based solutions/services. The cloud providers take care of ongoing maintenance and support of the baseline system and SaaS vendors do the same for their solutions/services, so you don't incur technical debt. Whether you decide to use an off-premise cloud-based solution/service or an on-premise one, focus on ways to eliminate technical debt.
- **Apply best practices to balance new and old technology** – iSpeak Cloud recommends this formula for resource allocation: 25 percent of resources allocated to hardening, 25 percent to new product introduction, 30 percent to unknowns and 20 percent to schedule padding. You can tweak these percentages depending on the velocity of the development team. The 30 percent allocated to unknowns covers items that were not planned for in an implementation as part of Agile as well as efforts to reduce technical debt such as end of life for legacy systems or consolidation of duplicate services.
- **Replace underperforming solutions** – Not all vendors can keep up with the changing needs of as-a-service solutions or environments. The CCoE needs to identify those vendors and ensure that the roadmap incorporates plans for addressing solutions/services from them. There may be open source alternatives that do a better job for a fraction of the cost.

In the end, the CCoE should evaluate, automate, adjust and eliminate technologies as needed to ensure the success of the cloud service platform.

At the end of this step you have completed your gap analysis related to people, process and technology.

Step 4: Align People, Process and Technology

In this step, the CCoE aligns people, processes and technologies with business objectives. Alignment is complicated by the human factor. Some projects are more politically charged than others.

Start with an impact analysis of tradeoffs based on the gap analysis performed in the previous step. This is one of the most difficult elements to build because many of the available off-the-shelf tools for analyzing and identifying data for the gap analysis do not always provide a proactive view. The task is challenging because solutions are either a custom-built patchwork of data pulls from disparate systems based on project versus lifecycle or too general to apply to the company's specific requirements. The best approach to resource planning is to integrate analytics tools with data sources for example, integrating ITSM tools with the financial planning system. Although creating an executive dashboard is optional, it's a huge assist for planning and implementing the cloud governance framework and the cloud service platform.

Distilling and identifying the right data to create a cohesive roadmap is not simple. The sheer number and velocity of the requirements of solutions/services can overwhelm even the best portfolio leader. One critical aspect is ensuring that the right level and type of data is clear, can be easily adjusted for tradeoffs and is accessible in a timely manner. For consistency across solutions/services for identification of patterns, costs and tradeoffs, iSpeak Cloud recommends using a common lexicon or terms to refer to them.
For example:

- **Business capabilities** – What business capability will the solution/service enable? It could be as simple as enabling electronic signatures on a SaaS solution or as complex as migrating service clusters across cloud providers to reduce costs. Remember business capabilities do not necessarily equate to services and vice versa. A business capability is a unit of functionality that enables the business in some new way to perform tasks.
- **Value** – The initial value may be intangible. At this point, the purpose is to agree on the value of each capability and/or supporting solution/service to the company. The end result is a ranking of business capabilities with respect to the value they will deliver—high, medium or low. Value can be based on cost savings, revenue generation or adherence to compliance and security requirements. Key stakeholders across the business who have cloud initiatives that align with company objectives should conduct these valuations.
- **Metrics** – What are the values or business case that the capability ties back to? Often, capabilities are double counted across solutions/services due to silos. For example, multiple business owners may create a separate business case for the same solution/service under a different initiative name, which can lead to significant inflation of the business case. Many leaders interviewed cited a SaaS CRM vendor that did a great job at selling the full number of subscription licenses needed for the entire company at the department level. Five of them reported having millions more invested in licenses than needed because they agreed to a subscription for the entire company multiple times for each separate project. Multiply that by hundreds, or in some large enterprise cases thousands of services, and you see how large the error can be. A major part of alignment is ensuring that the business cases are based on an accurate assessment of savings, revenue increase or cost avoidance based on centralized requirements.
- **Scale/Priority** – Where does the capability fit in terms of the core platform priority and the ability to scale out the platform to other service offerings? For example, a horizontal solution such as eSignature does not typically affect scaling the cloud service platform to a broader audience with additional services. On the other hand, failing to include monitoring or security will make scaling up or out difficult and costly.

- **Portfolio** – Solutions, services and microservices typically are bundled into a portfolio, initiative or program. (Ideally, at this point you have made the shift to a portfolio view as suggested in Phase 1 and also in ITPI's book *Visible Ops Private* Cloud. Business capabilities should be tied back to their portfolio for two reasons: to ensure the capability stakeholders are aware of the request and to identify and adjust for resource constraints with solutions in flight.
- **Dependencies** – The cloud service delivery manager records specific dependencies required for the given service across microservices. This includes integrations to other solutions, regulatory and compliance rules, resources, internal or external PaaS or SaaS calls, analytics and plugging into backend tools for monitoring and IT service management. For example, if the cloud service delivery manager is rolling out a container management solution to enable hybrid cloud bursting, there may be a dependency on the network access control (NAC) system to restrict access with respect to who can migrate data or applications, or to the operations orchestration tool to check the license database to be sure the rollout is legally permissible.
- **High-level estimates of time to value** – To determine when you can deliver or estimate a time to expect the solution/service to be delivered, you need high-level estimates. A "small, medium, large" approach lets you define the number of sprints (or weeks) that a feature will take to develop. The more integrations, the more people, the more dependencies, the bigger the size and the more time required to deliver the solution. Some Agile teams have come up with creative ways to do estimates so they do not have to be held accountable to velocity (number of hours) by using a point or reference system. Those methodologies cannot be easily measured and are often inaccurate. Velocity in terms of average hour per Scrum team, on the other hand, is a good indicator. Be sure to communicate with executives in terms they can relate to—for example, the impact on time to value, cost/savings and security.
- **Impacted groups** – What groups are typically affected by a new solution, new platform or new model? The CCoE needs to identify these groups early in the solution/service lifecycle and communicate with them regarding policy, strategy and process. For example, if the cloud service platform includes a self-healing service, the network operations center, tier-1 and tier-2 support teams and the change advisory board need to know about it and understand when it will be used.
- **Timing/release** – Timing and release is best displayed as a high-level calendar with the estimated rollout and final release time to value. For example, you may have one continuous integration and delivery release process for internal-facing solutions and a different one for external-facing solutions. Releasing a feature before its scheduled release date can have a devastating effect on revenue (for pricing and promotions) and on the technology team. Some in the organization will blame the initiative for the problem whereas the cause is actually human error due to lack of communication.

At the end of this step, you've completed a preliminary resource/velocity plan, which, combined with the BRD and gap analysis from previous steps, results in the roadmap to cloud.

Step 5: Define What Success Looks Like

Every pilot should begin and end with the fully scaled solution or service in mind. Before selecting a solution or service for a pilot, you must have a picture of the minimum viable setup in terms of functions required to ensure that the pilot is representative of the ultimate solution or service that will go into production. The CCoE needs to use both art and science to create a representative environment that is scalable to the production level and sustainable over the long haul. Success metrics should include both qualitative and quantitative measures that can be captured early, often and in an automated way.

Figure 15. **Embrace Digital Transformation Roadmap**

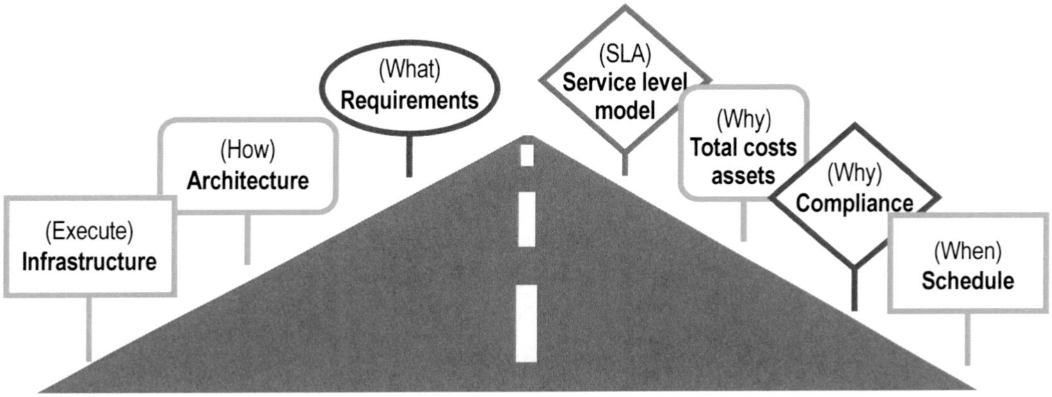

Success for the cloud service platform may look quite different than success for a typical solution-based offering. Top performers identified 10 common characteristics of a cloud service platform to ensure success:

1. **Standardize on cloud offerings (internal and third party)** – If possible, limit the number of cloud offerings people can choose to one or two external and one internal offering. Standardization reduces complexity, resource requirements and costs. The cloud service platform BRD should provide baseline requirements for cloud service providers and enable you to select the providers that meet the minimum viable product.
2. **Automate cloud service provisioning** – Develop an automated way to request, test, deploy, monitor and maintain third-party cloud provider services from a self-service portal. The automation should dynamically assign services to cloud offerings based on the security, cost and time-to-value policies specified in the BRD.
3. **Integrate with network access control and/or single sign-on** – Some form of role-based access control is essential, so develop a microservice that checks the user's role and, if that role has access to restricted data, limits the user's placement options to cloud solutions in the virtual private or private cloud.

4. **Automate change management** – Change management processes that are heavy handed can impede velocity and communication and exponentially increase duplication of data. You should automate or eliminate layers of change management for minor changes with minimal impact. This allows people to focus resources and time on critical changes that affect major systems, create risks or impact functions.
5. **Enable self service/assisted service** – Ensure that virtual machines, data, storage, microservices and other components of the cloud service platform can be requested, approved and received through a self-service or assisted-service portal in an effortless way.
6. **Provide day 0 support** – The cloud service platform must have some form of day 0 support and a clear escalation path that is defined, communicated and easy for the service desk to access through a standard knowledgebase or other mechanism. Simple requests such as resetting a password should be automated via a self-service portal or virtual assistant from day 0 to minimize calls into the service desk.
7. **Plan for business continuity/disaster recovery** – Cloud service providers will undoubtedly have failures. In extreme cases, a provider may cease doing business. Consequently, the core platform must provide a way to extract or replicate the affected solution/service, either to another provider or internally, with minimum disruption to any services that plug into the affected solution.
8. **Calculate chargeback/showback** – At a minimum, the cloud service platform must provide the ability to extract costs associated with a given service, department or individual. In addition, it's helpful to develop a microservice that automatically tracks costs by business unit, department or individual and triggers an alert if charges from that unit, department or person exceeds budgets. This helps keep spending visible and under control.
9. **Publish APIs (calls or injection)** – Every microservice on the platform should provide published and documented APIs that internal development teams and third-party solution providers can plug their solutions/services into. If this is not possible, some top performers provide an option for custom solutions to inject required code into the container or VM that hosts the solutions/services.
10. **Automate checks for security and compliance** – A two-component approach is recommended here. First, the security and compliance owner provides and inputs a list of rules required for segmentation, access control, data storage, minimum security patch levels and other elements. The rules are based on company and solution/service requirements around mandatory security and compliance controls. Second, the cloud service delivery team leverages an engine to apply the rules against a given solution or service. A solution/service that passes the screening is onboarded to the platform. One that fails the screening is flagged and items that need to be addressed are identified. This step accelerates scaling and onboarding of solutions/services by eliminating the need for extensive change oversight. It also reduces the introduction of risk due to human error.
11. **Provide dashboards for owners/leaders** – Create dashboards internally or use an out-of-box analytics reporting tool to present metrics that tie savings, costs and risks back to the business case. The metrics help with forecasting future projections of utilization.

Step 6: Identify Pilot Candidates

This step focuses on:

- Creating a list of solutions/services that align closely with company directives and requirements for technical, data, business and architectural fit
- Identify testing requirements for the proposed governance framework and cloud service platform
- Gathering useful information for scaling/extending the the cloud service platform

The best candidates are usually not as simple as email, nor are they as complex as unified communication. Be sure to select solutions/services that will fully test microservices and capabilities that must be part of the cloud service platform. For example, to test the platform's security microservices, you need to pilot at least one solution/service that requires security.

Be sure to select solutions/services that will fully test microservices and capabilities that must be part of the cloud service platform.

A good way to determine the fit of a solution/service for a pilot project is to create a matrix of the capabilities that clearly must be part of the cloud service platform—for example, security/compliance, eSignature and PDF reader. List the services and map them back to the components that will be tested during the pilot.

To ensure that you're getting a representative sampling of solutions and services that will stress test the viability of the platform, select services that have the following characteristics:

- A solid business case outlined and a BRD with supporting requirements
- Achievable and defensible return on investment with executive sponsorship
- Expandability to a larger audience, so if the pilot is successful it can be quickly ramped up and out
- Five integrations or more across various clouds (private and/or public), legacy systems and/or specialty applications
- Security and regulatory compliance requirements

Pick enough candidates to reduce the risks associated with the cancellation or delay of a project. In Phase 4, you'll take a more in-depth look at each pilot candidate and narrow the list down to three to five solutions/services to be piloted.

This step is instrumental in creating momentum to scale up and out as initial service offerings roll out. Remember, there are always issues with the introduction of any new solution or service.

Figure 16. **Summary of Phase 2—Roadmap to cloud**

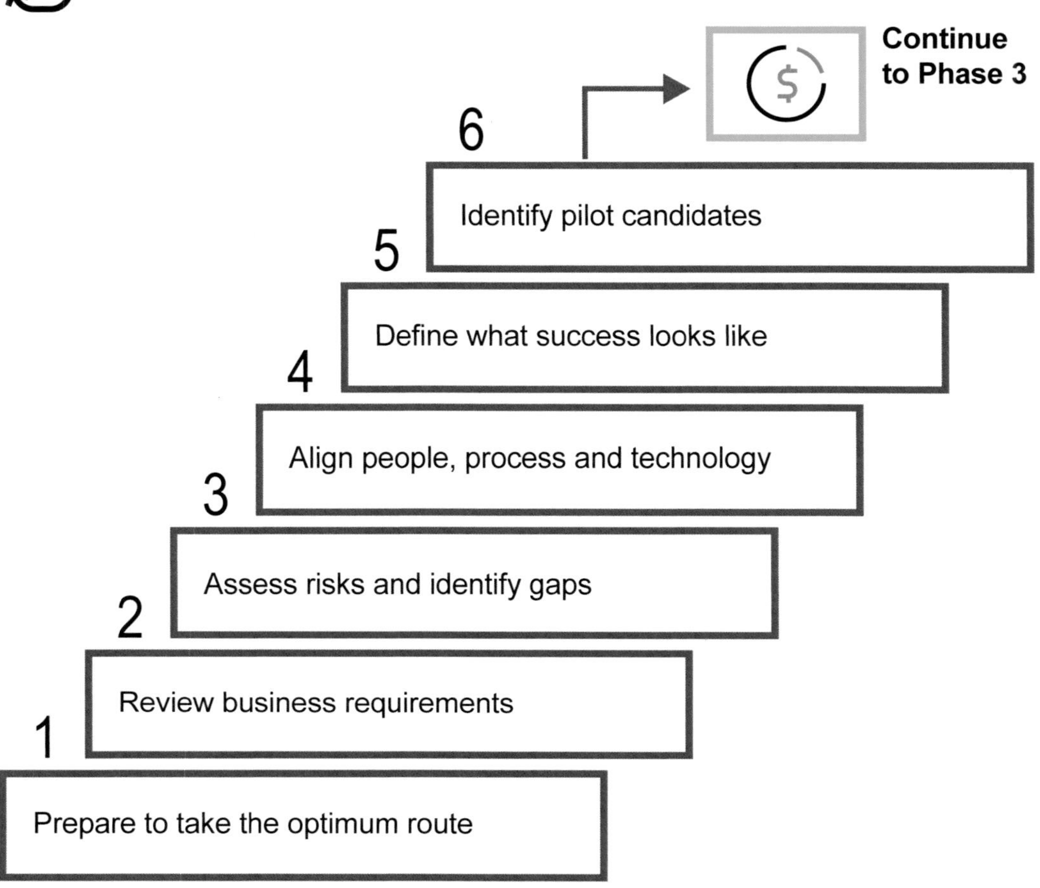

Words of Wisdom from the Trenches

Brian Elkins

Vice President of Engineering,
ExamSoft

Aligning People, Process and Technology

Persona Mapping and Retooling

It's not easy to get seasoned IT professionals to embrace the types of changes required by digital transformation in the everything-in-the-cloud era. One effective approach is to create persona profiles to educate cloud service delivery team members on who their customers are, what those customers expect from the solution/service and what the timelines are.

Mapping personas to the baseline cloud platform makes it easier to identify gaps and adjust the roadmap. A balanced roadmap for internal platforms versus external customer applications can get complicated unless you take the time to retool your resources. During the retooling it's vital to give people time to learn. Allocate sprint cycles for the research part of research and development. Slowing down in the beginning to focus on planning often enables people to speed up later while also making fewer mistakes when a solution/service rolls into production.

Centralize, then Push to the Edge

Regulatory restrictions for multinational customers can be challenging when trying to create a DevOps solution across hybrid clouds. It's critical to understand the multiple access demands of international customers. For those with data and privacy concerns, it helps to create centralized processes and microservices but then push the implementation out to the edge, where the data resides. For example, if the data resides in France or Germany because of privacy restrictions, IT would build the core microservices and supporting governance framework and cloud service platform to test the solution centrally with those solutions/services for which central testing is possible. Where central testing is not possible, IT may push the platform, framework and testing to the local source in an n-tier distribution model to the location of data origin. This centralize-then-push approach allows you to restrict regional solutions/services, data and microservices so they adhere to compliance directives.

Promote Your Roadmap Internally

Depending on IT's reputation among business users, it may be very easy or very hard to move up the continuum to the business partner or trusted advisor level. For this reason, technology leaders owe it to their people to not only manage up and down but also to market their solutions/services internally. If you show executives the savings, effectiveness of the program and positive impact, they'll have a hard time arguing when you ask for more funding to transition from pilot to production.

PHASE 3

DETERMINE CLOUD COSTS

"Third-party cloud solutions are not always the least expensive. They do, however, guarantee less technical debt because the providers have to update their systems on a more common schedule to meet service level agreements. It's a lot easier to have a vendor enhance an application to address changing customer needs than to fight the internal battles for years to get funding for upgrading systems."

—Aaron Amendolia
VP Service Delivery

Phase 3 Objectives

Develop a solid understanding of all costs, challenges and end-of-life considerations needed to make an informed decision with respect to the cloud strategy.

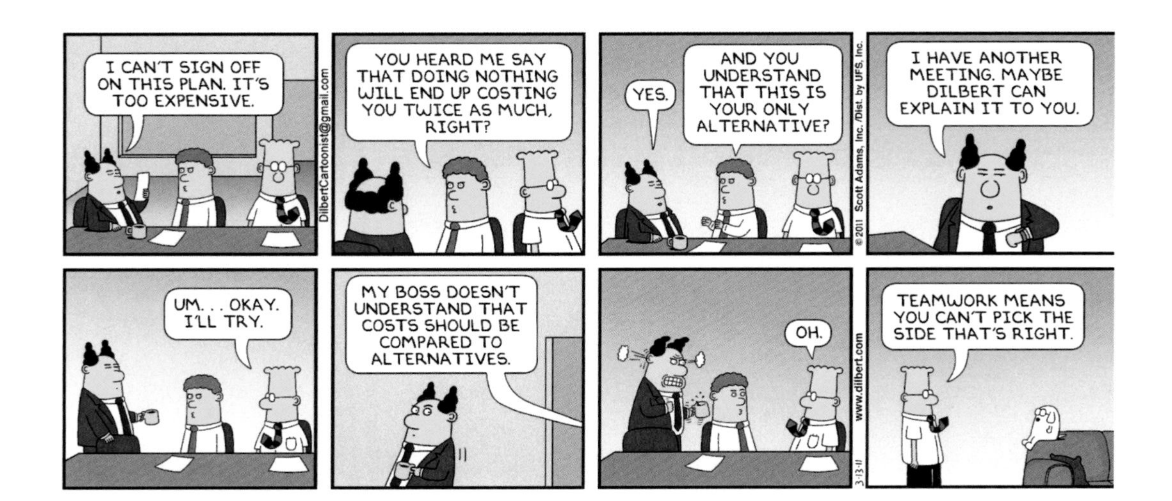

Issues and Clues

The following table summarizes issues and clues that Phase 3 addresses.

Issues	Clues
Lack of Visibility to TCO	"Agile adds velocity but clouds visibility for costs. How do we project a financial plan when we cannot measure hours or unit costs?" –Financial Planning and Analysis "Our ISP costs skyrocketed. We were 4.5 times over our monthly budget because of Shadow IT solutions. We identified over 300 that were unknown to IT during an audit." –VP Operations
Not tying back to a business case	"Our best guess approach has led to many critical projects like self service being delayed while other less critical pet projects received the resources and focus. Portfolio balancing becomes very political if it can't be tied back to real numbers." –Business Leader "SaaS solutions and other providers are overselling capabilities to my peers. After the contracts go away the technology team is left trying to figure out how to make up for areas where the system falls short without the resources or budget." –CIO

Issues	Clues
Lack of Collaboration	"The business purchased the solution and now we are asked to implement it without any budget. The business case did not include critical elements like the messaging service or additional integrations needed. Now we have to go back and ask for more budget." –Cloud Service Delivery Manager "The SaaS vendor successfully sold the maximum number of licenses needed to every line of business owner. Their desire to own the solution just meant wasting millions of dollars on licenses we will not use." –CIO
Using the wrong model	"Hybrid cloud solutions are the gift that just keeps on giving. The three-year model provided by the SaaS vendor appeared to be less until we created all the support infrastructure. In the end it was 30% more than the on premise solution we have." –CIO "Business cases are built to cover a three-year period and are tracked only by the project. The issue is the services live on after the 3-years in many cases and there are no projects to reduce the technical debt. More often than not the company is put at risk." –Enterprise Architect
Not forecasting growth	"Our program manager decided to place the project on Amazon because it was easier than getting the resources from the operations team. At first it seemed like a great idea until we rolled out the solution at scale. Further analysis showed we would cut our costs in half by running it on a private cloud." –Enterprise Architect

A Tricky but Critical Task

In this phase, you will undertake a critical task: estimating the costs of the initiative and correlating the costs to compliance with company directives. The task is especially challenging in that it is involves both science and art. Consequently, it's a task that is often not well executed. The biggest challenges are understanding what metrics to measure from the beginning, collecting the right data to gauge those metrics, and tying the metrics back to a business case.

You will tie together three pieces that, when combined, provide cost visibility from the solution/service at the top of the stack down to the underlying platform. These pieces are:

- **Business requirements document** – The BRD created in Phase 2 ties together not only the services but also the business capabilities from plans back to the underlying technology. You need a BRD for each solution/service under consideration for pilot, one for the cloud service platform and possibly one for each competency center.

- **Profit and loss (P&L) cost analysis** – Unlike a cash flow analysis, the P&L analysis takes into account the depreciation benefits the company receives from its on-premise/legacy systems and resources.
- **Master planning (optional)** – A master planning dashboard or other tool is essential for consolidating data on resources, requirements and timeframes for each solution/service as well as for the cloud service platform. The primary purpose of this tool is to tie back the detailed planning (user stories), resources and project schedules to the BRD (business capabilities/epics) and cost analysis.

If the CCoE does a good job in this phase, the company can attain high levels of agility and performance in bringing new services to the cloud platform while coming in at projected cost levels. That can create momentum for adding more services to the platform.

The following tips will help you avoid some of the most common mistakes companies make in determining the true cost of solutions and services:

- Throw out assumptions – People sometimes make assumptions about the costs and savings of a particular solution. Validate all assumptions and be thorough in identifying cost contributors and in assessing how much the solution or service will save.
- Use a P&L model – Cash flow models work well when software and hardware are perpetual or owned versus rented. In this traditional approach to costing, being slightly off with your cost/benefit analysis isn't too serious because there is a margin of error built in for depreciation benefits and longer product lifespans. Cloud-based solutions, however, change the economic picture because of the rental or subscription-based model in which the company loses depreciation benefits. To accurately compare costs among various cloud-based options or retaining existing implementation, you have to do a P&L analysis that takes depreciation into account.
- Distinguish between the solution/service and the platform –The cloud service platform BRD should clearly state the savings and benefits it will deliver. The business cases and BRDs for the solutions and services should also state the savings and benefits they will deliver. Make sure that solution/service business cases and BRDs do not include benefits and savings that will be delivered by the platform. Doing so will result in an overstatement of savings. If you include platform savings in five separate solution/service business cases and BRDs, the savings are overstated by four times.
- Plan based on lifecycle – In too many companies, once a solution is implemented, the project is deemed completed. In reality, resources are needed beyond the launch for ongoing maintenance and support of the solution/service. Failing to account for ongoing costs associated with the complete lifecycle increases technical debt and places a burden on employees trying to do the right thing. The cost of supporting a service, keeping it updated and incorporating enhancements often far exceeds the initial development costs.

Step 1: Create Next-level Capability Mapping

In this step, the CCoE turns business capabilities into epics (or high-level features if you use waterfall) for the services that were selected for the pilot.

Figure 17 depicts an illustration of a simplified mapping between the high-level business capabilities from the BRD such as electronic signatures with additional details from the master planning solution (requirements database). This high-level mapping enables tying the business capability cost back to the features. It also enables identifying patterns of services used repeatedly that can be consolidated into a microservice.

Figure 17. **Mapping of business capability to the epics that drive the solution and the cloud platform.**

Business capability: The electronic signatures capability will enable customers to submit requests, contracts and orders electronically.

Epics: A single capability could encompass multiple high-level functionality or epics and touch multiple products or services.

Author	Marissa Sato	Marissa Sato	Yoda Knight
1=Critical, 2=High, 3=Med, 4=Low	1	1	2
Business capability	110	110	110
Business capability	Electronic signatures	Electronic signatures	Electronic signatures
Product	Online Banking	Audit	Security
Epic (feature)	Electronic signature for wire transfer	Audit of electronic signature for wire transfer	Secure electronic signature for wire transfer
User story - One-sentence description "(User or system) should be able to __________ so that they can __________"	The Online Banking system is to display icon representing electronic signature for the user when the statement feature is clicked.	The customer database will provide audit trail capabilities for the machine learning program used by audit to detect fraud.	The electronic signature solution will be encrypted at rest and transport to protect personally identified information.

Author	Marissa Sato	Marissa Sato	Yoda Knight
Detailed description (sub-requirements)	The system will prompt the user to read the disclosures prior to allowing them to sign the documents.	The machine learning program will analyze patterns of signatures and alert auditors of anomalies.	The system will apply standard encryption for personally identifiable information in the loan and submission process.
Author	Marissa Sato	Marissa Sato	Yoda Knight
Acceptance criteria	When the user clicks on the I acknowledge reading the disclosures, the system will provide the ability to submit the signature and the user to download a receipt.	When an anomaly is detected the auditor will receive a prompt from the machine learning system.	When auditor tries to access data it is obfuscated.
Dependencies/ assumptions	Adobe PDF reader, search capability	Machine learning program	Data encryption
Service level	Service level agreement of 98.9% or silver	Service level agreement of 99.9% or gold	Service level agreement of 99.9% or gold
Exists today	N	N	N
Roadmap	Y	Y	N
Time to value	9/1/16	9/1/16	9/1/16
Product manager	Carolyn Zarate	Ashley Sticca	Linda Kavanaugh

The epics in Figure 17 show that the capability must integrate with multiple services or products, raising the requirement for cross-functional collaboration. Consequently, this electronic signatures capability is a good candidate for implementation at the platform level in the form of a microservice that each product or service can call using APIs. Developing the microservice once and sharing it across multiple products and services results in higher development efficiency and agility, and lower costs.

Step 2: Identify the Cost of Additional Integrations and Services

In this step, you account for costs associated with integrations and requirements that may not have been specified in the original business case or in the high-level cost estimates used to create the roadmap in Phase 2. As teams create the detailed user stories to accompany the epics and business capabilities, they may cover additional units of work that must be performed for the microservice to be functional. Or there may be additional hidden costs or costs associated with third parties that were previously unknown.

For example, the last epic in Figure 17 is not listed in the roadmap. However, it is critical for the service. The epic must be vetted like any other newly discovered item. If it does not fit into the resource, time or security requirements, the CCoE must determine the next best course of action. Another example is the epic in Figure 18, which shows that the next-level requirements for the epic are only partially met. Therefore, additional work must be scoped to complete the epic and minimum viable service for the business capability to realize the ROI stated in the business case. The additional expenses (resources, software, time and others) must be forecasted and added to the business case so that true costs are captured.

Figure 18. **An example of a user story identifying a partial solution that may have additional cost implications to complete.**

Author	Mickey Pluto	Ginny Potter
1=Critical, 2=High, 3=Med, 4=Low	2	3
Business capability	110	110
Business capability	Electronic signatures	Electronic signatures
Product	Omni-Channel	Omni-Channel
Epic (feature)	Secure electronic signature for wire transfer on mobile devices	Secure electronic signature for wire transfer on from website
User story - One-sentence description "(User or system) should be able to _________ so that they can ___________"	The Mobile Banking application on iOS and Android will have an icon to enable electronic signatures.	The Bank4U Website will allow customers to upload electronically signed applications from SMS, email, or computer.
Detailed description (sub-requirements)	The system will provide the same icon a as the current desktop and future web application.	The system will allow uploads from various communication channels.

Author	Mickey Pluto	Ginny Potter
Acceptance criteria	When the user clicks on the icon from on either IOS or Android they are walked through e-signature prompts.	When user clicks selects their signature file they can extract it from SMS, email or computer.
Dependencies/ assumptions	Mobile banking application on iOS/Android	Upload capabiltiies for PDF from Bank4U Website
Service level	Service level agreement of 95.9% or copper	Service level agreement of 98.9% or silver
Exists today	N	Partial
Roadmap	Y	Y
Time to value	10/15/16	11/2/16
Product manager	Mark Bodman	Casey Carlson

This step also includes the optimization of service levels across integrated solutions. Too often portfolio managers or business product managers define a service level agreement an SLA based on the primary applications or microservices it applies to or the new applications being created as part of the offering. What is often overlooked is the cost to uplift legacy applications or data that the service will consume.

Notice that the service levels in Figures 17 and 18 vary from gold to copper. Only two of the services require gold-level performance, yet the microservice is specified to perform at the gold level for all services. That unnecessarily increases costs for the services requiring only silver and copper service levels. In cases such as this, the CCoE should work with the product managers to determine which services require only silver-level performance. For those services that do require gold performance, the CCoE might request that the additional costs associated with the gold service level be incorporated into the budgets for those services.

In this step, the cloud service delivery manager works with the solution and service owners who created BRDs to ensure that all essential elements of the solution or service are included. Assisting owners in the effort to identify all such omissions ensures that the project is accurately scoped and that there won't be any budget-busting cost surprises down the road.

Finally, the CCoE works with the cloud service delivery manager to determine when the solution or service must be available to the cloud service platform team to ensure that the team has a sufficient number of development cycles to roll the service into production by its required time-to-market deadline. This task also helps surface resource hotspots and cold spots. For example, a seasoned director of development can look at the chart shown in Figures 17 and 18 and see that the development team will be very busy in the July through November timeframe.

Another important task in this step is to ensure that the people who created the platform are available to support the rollout in case there are issues with the integrations, scalability, security or other requirements specified in Phase 2.

Step 3: Add Scalability Costs

Often, a pilot goes extremely well but the piloted solutions or services begin to crumble when they are rolled out to tens of thousands or perhaps millions of users. This step ensures that everyone looks beyond the pilot to understand the minimum requirements for a large-scale rollout. It also ensures that the CCoE captures scaling costs and decides how to charge back or show back scalability costs. This approach helps secure funding for the cloud services platform. It also identifies risks to SLAs, services and the enterprise overall due to scaling. For example, at a large education software company, the enterprise architects did not take into account the fact that adding a new solution that included a third-party cloud service would increase the number of "trips" through the enterprise network to the cloud provider's environment. The sharp increase in network traffic burned out the primary network switch and, in time, the backup switch as well, taking the network down for nearly three days. The new solution had to be rolled back and users had to revert to using the previous on-premise solution until the team could determine requirements for upgrading the networking infrastructure and refactor the service to achieve the required scalability. Asking the right questions is essential to understanding the cost of scalability, as outlined below.

Infrastructure

To determine the infrastructure costs associated with scaling, the CCoE needs to answer questions such as:

- How many users will be accessing the service?
- What is the average number of transactions per user during peak hours and data transfers per transaction?
- What is the average load or consumption per user?
- What is the maximum number of users that can be supported by a VM from the vendor/enterprise architects for the service?
- Are there any pilot services/numbers that can be accessed? If so, who are the providers and what are the initial costs for:
    - The type of VM to be leveraged (What are the cost differentiators?)
    - Restrictions (such as licensing) that require a certain type of VM license
    - Each VM instance
    - Monitoring
    - Storage, disaster recovery/business continuity or maintenance/support from a cloud service provider
- Are volume discounts available from the provider? If so, what are the thresholds negotiated?
- Can the service be combined with another one to qualify for higher discounts from the cloud service provider?

ISP and Network

To determine ISP and network costs, the CCoE needs to answer questions such as:

- What will the increase in load be? Can you forecast future load?
- Can the current networking infrastructure handle the load increase or are upgrades needed for networking, capacity, security and other areas?
- Do you need to upgrade or renegotiate the current ISP contract?

Client Systems

It may not seem intuitive for client questions to be a part of scalability for cloud solutions, but there are a few items in the client systems and/or configurations that could affect scalability of the cloud service platform. Specific questions to ask include:

- Do the standard client configurations meet the hardware, operating system, connectivity and plugin configurations needed for the pilot services?
- Are upgrades needed? If so, what upgrades and in what timeframe?
- Are there additional costs or performance concerns with the service being accessed by mobile devices or virtual desktops? If so, what are they?

Step 4: Factor in Remaining Costs

Be sure to include any remaining costs that have not been addressed in the preceding steps of this phase.

Training, Documentation, Organizational Change and Development

For costs associated with this area, ask such questions as:

- Do the users require documentation and/or training on new processes, systems or environment? If so, which users and in what timeframe?
- What resources are needed to conduct the training?
- Is new content required—for example, video clips and knowledgebase articles?
- Are there legal or other requirements for training and/or sign off?
- Are there new reporting applications or systems required by organizational change development to track and monitor the effectiveness of the change?

The costs for this area vary based the position of the service in its lifecycle. New services will require more training and documentation than updates to current ones. The intuitiveness of the user interface and the technical maturity of the user community will also affect costs in this area.

Marketing and Communications

Publicizing your wins internally and, if applicable, externally are important in obtaining broad adoption of the solution. So be sure to include the costs of these activities in the budget. The marketing and communications plan should include ways to measure and communicate success using key performance indicators (KPIs). The CCoE should distribute regular success updates to users, executive sponsors and employees to ensure that people focus on successes instead of on what might be going wrong.

You can accomplish a lot with just a few resources. For example, with just one writer and one communications director, a large university clearly communicated the intent and successes of a transformation initiative to 50,000 affected employees. These two people published documentation that described the transformation and knowledgebase articles that the employees could access for additional information.

Security and Compliance

The cloud service delivery manager will work with the security strategist to refine the cost, controls and requirements from the BRD defined in Phase 1. All the associated costs and requirements should align at a minimum with the service requirements for security and compliance. Any additional costs should be tied back to the appropriate service business case.

Performance Monitoring

Monitoring performance in hybrid cloud environments is complicated, so it helps to develop tools or leverage existing ones that provide data and analytics capabilities to support performance monitoring. For enterprises that don't have these tools or don't have budget to purchase them, open source tools are available. However, open source tools may not provide sufficient information to support effective troubleshooting.

A global manufacturing firm created a microservice that monitors the performance of a system that integrates the company's on-premise telephony and financial records with a SaaS CRM solution. The microservice was needed because there were no APIs that allowed integration of the company's monitoring tool with the CRM solution, and, therefore, no way to determine if the CRM solution was meeting availability and performance requirements.

Business Continuity and Support for Hybrid Models

Another often-overlooked item is a day 0 business continuity and support plan for solutions/services running in hybrid cloud environments. Be sure to include the following in your cost estimates:

- **Multiple cloud providers involved in backup plan** – A sound backup plan includes at least two cloud providers (internal and external) with backup to the secondary provider. This approach avoids situations such as that faced by Verizon customers when that company announced it would no longer provide a third-party cloud platform.[15] Customers had two months to move or lose their applications and data.

A sound backup plan includes at least two cloud providers (internal and external) with backup to the secondary provider.

- **Outages** – With every cloud, there is the potential for large-scale outages. Your backup plan should address what happens when outages occur and document a clear recovery process. Consider defining such capabilities as caching data on premise to enable continuity of service and/or running a daily on-premise backup with failover to an internal private cloud. Costs related to the backup plan should be part of the budget for the cloud service platform and possibly charged back to the solutions/services. Pay particular attention to handling public cloud outages.
- **Troubleshooting and resolution** – When a service comprises many different components, finding and resolving issues can be difficult. A large educational services company integrated 48 applications and microservices into a single higher-level service. The first time the higher-level service went down, troubleshooting and resolution took nearly three days because the problem involved four components that were conflicting with each other. IT had to figure out the root cause, fix the integrations and then build out microservices with the assistance of the cloud and SaaS providers to eliminate conflicts between the solutions. Provide funding for tools and automations that accelerate troubleshooting and resolution, including:
  - Analytics tools to review changes and service maps quickly
  - A self-healing capability that automates rollback and roll forward of microservices
  - Automated change record update
  - Vendor tools for troubleshooting, including log file analytics, error messages from API calls and other items noted from vendors or the open source community

---

15. "Verizon to Close Two Public Cloud Services," Barb Darrow, February 2016. http://fortune.com/2016/02/12/verizon-closes-two-public-cloud-services/

Step 5: Create a Cost Analysis across Platform Services

The financial planning and analysis and fixed asset accounting team members are front and center in the conversation for this step. The step involves costing out three to five services. Before you can do that, ensure that each solution/service owner adjusts his or her business case based on the costing information presented in Steps 1 through 4 of this phase.

While you are building the cost analysis there may be future services for a given portfolio of services that can share or utilize a microservice that is being created. When this occurs you must work with the portfolio owner to build the initial costs in the first services and then provide some way for showback as additional services are onboarded from that portfolio owner.

You need a standard cost structure for solutions, services and the platform so you can compare apples to apples. Each solution/service should have a business case that is based on the net present value (NPV) for the depreciable life of the solution/service regardless of whether it is a SaaS or on-premise solution/service.

For example, if you're evaluating a CRM solution with a five-to-seven-year lifespan, then the business case should cover the entire seven years you will have to pay the subscription. Fixed asset accounting can provide the timeframes the company uses for depreciable life of on-premise services.

Just because a service resides in the cloud doesn't mean that the company won't use it for as many years as a traditional solution. Nor does it mean that it will take less time to implement. Whether the solution runs in the cloud or on premise, a certain amount of time will be required to gather requirements, change processes and roll out the solution. Companies are often caught off guard when it takes longer to implement a service in a hybrid cloud environment than it would have taken to implement on premise. More often than not the cost to implement a hybrid cloud solution is more expensive in some areas and less in others. The key is to weigh the risks, costs and long term ramifications before deciding.

Just because a service resides in the cloud doesn't mean that the company won't use it for as many years as a traditional solution.

Standardize the Model across Services

To ensure an apples-to-apples comparison of costs for the solutions/services you're evaluating, you need a common method for chargeback and showback. The model must be granular to show which items and costs can be shared across services and which items should be charged to the relevant business groups based on consumption. Consumption models may hit a threshold where the costs exceed the benefits. For this reason it is important to forecast and understand at what point the costs start to outweigh th benefits.

Cost and license models have implications not only for people and process but also for architecture. Several factors can affect the company's standard cost model. One factor is how granular the model is. The politics around chargeback and showback affect the entire stack. The heaviest consumers are typically the ones to push back the hardest on chargeback and showback policies.

In the beginning you have to determine, based on the services, whether a chargeback or showback model will work for your company. You have to decide which one you're going to use and then figure out the level of granularity you need and how you're going to capture the costing information so you can charge back or show back the costs. Some detailed questions to ask during this determination:

- What detail layers in the stack are being charged back for?
- Does the company want a breakdown of the VM instances, data, storage, networking, and ISP, or a consolidated view of all of them?
- Are there hierarchical layers or multitenant costs that are expected to be charged back? For example, if there are many business units or separate entities that share a service in a multitenant environment, how would accounting like the costs to be broken down?
- What checks and balances do you have with procurement to automate the legal and technical ramifications of licensing and enforcing the existing agreements?

The CCoE needs to create a comparison chart for chargeback like the one shown in Figure 19. This should be derived across the current cloud service providers the company is using or considering. There should be some references from current services using private, public or hybrid cloud infrastructures. This comparison helps determine where to place workloads based on thresholds/limits of providers and provides a common language with respect to features and functionality. The CCoE uses the chart to:

- Identify optimal architecture (private, third party or hybrid) for each solution/service
- Identify tipping point where a service could no longer be cost effective because of time, security or other aspects on third-party clouds
- Discuss exceptions such as innovation centers or small incubators to insulate and isolate new solution ideas
- Provide a common language/agreement across business cases to apply to new services or legacy service migrations
- Identify additional costs and technical debt for tagging to track costs and licensing
- Demonstrate how cloud costs are all over the map and show the value of consolidation to reduce shadow IT
- Determine the best routes to cloud value

Figure 19 illustrates an initial chargeback model to help service owners beef up their business cases to justify the cloud service platform. Interestingly, the hybrid model is the most costly because of the additional requirements for caching data for business continuity, integrations between multiple cloud providers, additional facilities requirements and support.

Figure 19. **Chargeback comparison chart**

Cost to run per month	Hybrid	Private cloud	Third-party cloud
Instance cost per Unit	$19.09	$27.31	$54.84
Storage	$8.44	Included	$5.89
Data transfer	N/A	$5.38	$2.08
Disaster recovery	$4.77	Included	$13.71
Labor (Ops VM staff)	$10.40	$10.40	$10.40
Security (Anti-virus, patch licenses)	$7.34	$2.45	Included
Network operations center/ security labor (cost of full time engineers)	$2.31	$2.31	Included
Facilities (power/cooling - est.)	$104.36	$6.50	Included
Total	$156.71	$54.35	$86.92

Note: Hybrid = Legacy, Private, 3rd party. Private Cloud is strictly private cloud implementation, third party would be solely on third party provider. This is an example of a single service comparison. The actual VM cost comes from projected price quotes based on volume (number of instances), size of VM, licensing and other equal elements that can be compared. When you hit higher volumes, the law of diminishing returns comes into play. This comparison is used to optimize route to cloud value for the given service based on negotiated rates from company or current costs as is the case with legacy or hybrid systems.

Disaster Recovery estimated @ 25% of instance costs.

> Hint: Ask CCoE participants to bring examples of cloud provider current contracts or quotes for a given service.

Do a sanity check of your comparisons across providers. It may appear that a startup cloud provider offers more attractive pricing for a service than Amazon AWS. You may find, however, that a new cloud service provider costs less because it doesn't have the global coverage, security or reliability of an established player like Amazon.

Map Requirements to Costs and Benefits

Based on the business cases for the services, the team can build out the platform business case. First, the CCoE members map the high-level business requirements back to projected costs based on the standards created across providers, microservices, and required scale. This includes calling out integrations, resources and infrastructure for private cloud solutions, networking and other infrastructure needs.

Your goal is to determine hot spots, cold spots and true costs as well so you can narrow down your list of candidates for piloting the cloud service platform. The cloud service delivery manager should drive the integration, tuning and timing discussions with the solution/service owners, enterprise architects and the cloud service delivery team. Discussions should cover consolidation and benefits related to resources, end-of-life strategy and cloud service providers.

Figure 20. **Chargeback comparison chart**
The CCoE must balance the costs and benefits to the company as part of their process for determining the right mix of services and strategy based on budgets.

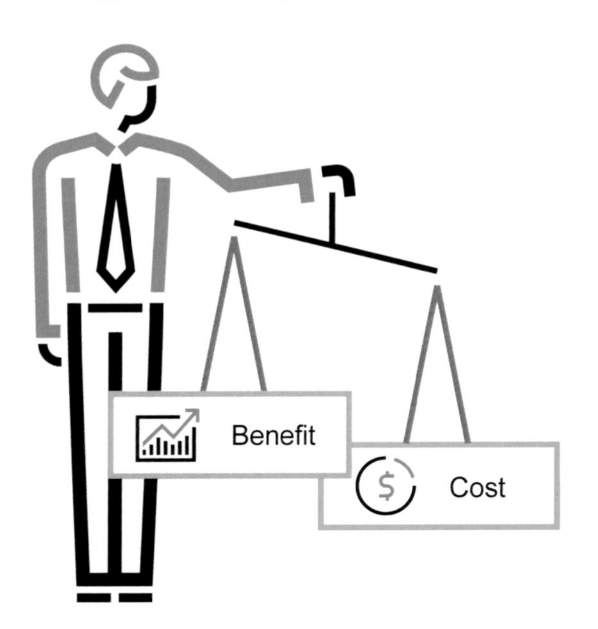

Resources

Identify human resources, processes and technologies that can be consolidated into the platform. Examples might include multiple CMDBs and reporting tools as well as people who can be reallocated from security, operations, integrations or development functions to strategic roles such as creating microservices.

Figure out the benefits of addressing hot spots and cold spots and consolidating resources, processes and technologies. Consolidating multiple tools with a single tool will likely result in cost savings. For example, consolidating multiple tools under one hybrid cloud management tool can reduce the maintenance, various required skill sets, and time orchestrate amongst tools.

End-of-life Strategy

Determine what you can eliminate or repurpose as a result of rolling out the cloud service platform and its supporting microservices. If the platform includes an eSignature capability as a reusable microservice, how much does the enterprise save over incorporating it into individual solutions/services? Moreover, can you streamline the current process for obtaining signatures. For example, if the current eSignature process includes eFax because it the past it was a requirement for getting electronic signatures, figure out if it is still an essential capability. If not, determine the timeframe for phasing it out and calculate how much the company saves as a result. Also identify any benefits from repurposing people or hardware once components are replaced.

Accept SaaS can have CapEx

Although SaaS solutions do not require the same amount of infrastructure and resources as an on-premise solution, there are still resource and infrastructure costs. Figure 21 shows that even a SaaS solution can include CapEx. In the figure, the cost of developing integrations into legacy solutions fall under CapEx. The cost of the human resources building those integrations would also be CapEx. Resources working on customizations to the SaaS solution, however, would likely be classified as OpEx. The financial planning and analysis members of the CCoE can provide insight into the appropriate methods for determining which efforts are CapEx and which are OpEx, providing insight into which ones are depreciable.

Figure 21. **Cost Analysis**

Initiative	2014	2015	2016	2017	2018	TCO
CRM (OpEx)	$1,000K	$1,000K	$1,000K	$1,000K	$1,000K	$5,000K
Integration (CapEx)	$1,600K	$0	$0	$0	$0	$1,600K
Integration (OpEx)	$400K	$50K	$50K	$50K	$50K	$600K
Resources (OpEx)	$200K	$200K	$100K	$100K	$100K	$700K
Resources (CapEx)	$800K	$800K	$0	$0	$0	$1,600K
Infrastr (CapEx)	$800K	$0	$0	$0	$0	$800K
Infrastr (OpEx)		$150K	$150K	$150K	$150K	$600K
Integration (D)	-$320K	-$320K	-$320K	-$320K	-$320K	-$1,600K
Resources (D)	-$320K	-$320K	-$320K	-$320K	-$320K	-$1,600K
Infrastr (D)	-$160K	-$160K	-$160K	-$160K	-$160K	-$800K
Total						$6,900K

Figure 22. **Summary of Phase 3—Determine cloud costs and compliance**

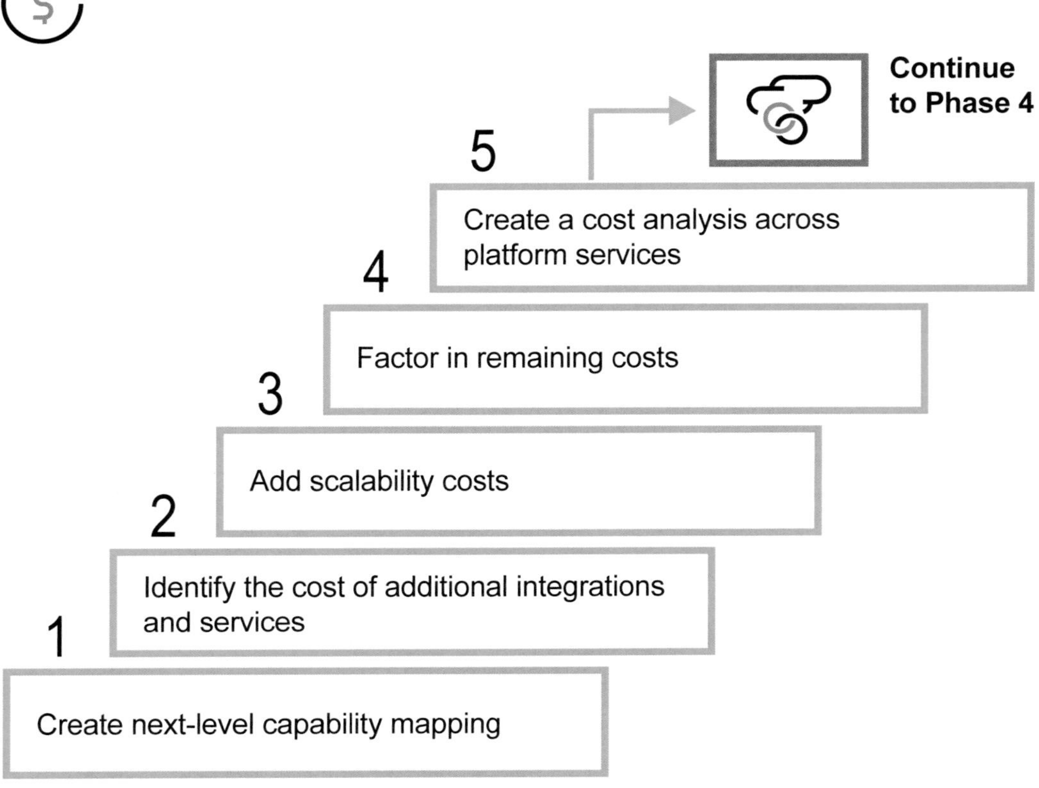

Words of Wisdom from the Trenches

Brian Cinque
Cloud Service Delivery Manager,
GE

Billing Models for Cloud Deployments

Chargeback or Showback

The cloud service delivery manager should start by determining if management supports a chargeback (billing) model or a showback model. Management's preference affects budgeting and funding and has technical implications related to capturing and reporting on costs. For example, for fine-grained billing, let's say at the application level, technical planning and development must go all the way down to tagging and tracking each application.

Customer First

Gain insight into the customer's perception of chargeback, especially if customers aren't currently charged for the resources they consume. Find out if business units plan to add any charges to the corporate charges before presenting them to their customers. Doing so could have technical implications for the tagging used by the cloud service provider.

Avoid customer pushback by clearly communicating why you're implementing chargeback and how it works. Otherwise, you'll experience "death by a thousand pecks," and you'll spend an inordinate amount of time chasing down details for owners who want to understand costs and contain them. Put your customer first by using a simple-to-understand chargeback approach. Avoid à la carte and custom solutions and instead create a consolidated list of features and costs. Keep the people who will request the costing data in mind when you develop presentation formats. Standardize reports and views and align them with roles such as product owner and financial planning and analysis, and then tailor data presentations for each audience.

Limitations and Standards

Work with the finance and technical communities around the tag standard they are willing to implement and use. If people don't buy in, you won't have a consistent way to track charges or usage. Consider using third-party financial reporting solutions that provide a self-service portal for application owners so they can get answers to basic costing questions. As you investigate these solutions, ensure that the finance team is heavily involved—even driving the process. Typically, financial teams look at the costing models in a general ledger format that complies with accounting standards. Help them understand that the data they need is there and that it's simply presented in a different way.

PHASE 4

CALIBRATE CLOUD VISION TO REALITY

"Everyone in the company must understand that technology and the business are all part of the same team. Technology leaders must embrace change to build a partnership of trust and have a seat at the table with their peers. Leaders across the company should work together with IT to automate legacy processes that hinder progress."

—William Velez
CIO, International Financial Service Firm

Phase 4 Objectives

Support the cloud service delivery team's efforts in designing and building a cloud service platform capable of supporting the pilot solutions/services and beyond.

Issues and Clues

The following table lists issues the CCoE addresses in Phase 4 and provides actual quotes from IT leaders interviewed for this book.

Issues	Clues
Cloud Sprawl	"Our yearly audit uncovered 300 cloud-based applications that IT wasn't aware of. It took months to track down the owners." –Cloud Security Architect "We cut AWS costs 30% after working with business and IT owners to consolidate cloud initiatives." –Cloud Architect
Broken processes/ Lack of trust in IT	"We mandated that the business get IT approval for SaaS solutions. The business ignored us and acquired 100 more SaaS solutions without our input." –IT Director "Our process for acquiring cloud technology is dysfunctional. One guy requested software so he could work from home while recovering from an injury. He recovered and returned to work long before he got the approval for the software." –IT Director
Failure to Launch	"The VP of one of our business units hired a third party to develop a custom application. Three years later the third party and funding are gone. Now IT is supposed to enhance the application without resources or funding." –CIO "Over three years, our innovation officer presented 78 ideas for transforming the company to a digital enterprise. Not one was implemented successfully." –Executive Program Manager

Issues	Clues
Cloud Cost Surprises	"The business owner of a large SaaS solution overstated benefits by five times and understated costs and time by 50%. We would never have approved the project if we had known the real costs." –Business Leader "Our CFO pulled line items from expense reports and discovered we're spending $9 million a month more than projected on cloud providers. We created a policy prohibiting the expensing of VMs and forcing people to direct requests through IT where we can centrally managed them." –Director, Cloud Service Delivery
Duplicate Tools	"We had 37 different applications for entering a customer address. None of them were connected. Customers were upset because they would change the address in one place and expect the change to propagate everywhere else." –Director, Service Desk "We have multiple tools for the same job. Developers can choose their own tools, which increases complexity. We had to come up with a consolidation project." –Chief Product Officer

Step 1: Design, Build and Test the Cloud Service Platform

The term cloud means different things to different people. To avoid confusion, as stated previously, iSpeak Cloud uses the NIST definition for cloud.[16]

The first step in this phase is to review documents created in earlier phases to ensure that they are complete and that they accurately define the minimum set of features and functions the cloud service platform must have to support pilot solutions and services.

Architect the Cloud Service Delivery Platform

Under the direction of the cloud service delivery manager, the cloud service delivery team designs the architecture for the platform. The cloud service delivery manager ensures the platform is properly vetted with enterprise and infrastructure architects and presents it to the CCoE for review and approval.

Figure 23 shows a conceptual architecture. At the top, you see the solutions and services that will be delivered through the cloud. The middle layer provides the management fabric for orchestration, abstraction and automation. The bottom layer is the foundation containing the microservices and functionality required for the first few platform releases.

16. The National Institute of Standards and Technology definition states that a cloud solution must, at a minimum, possess five characteristics: on-demand self service, broad network access, resource pooling, rapid elasticity and measurability.

Figure 23. **Cloud service platform conceptual architecture**

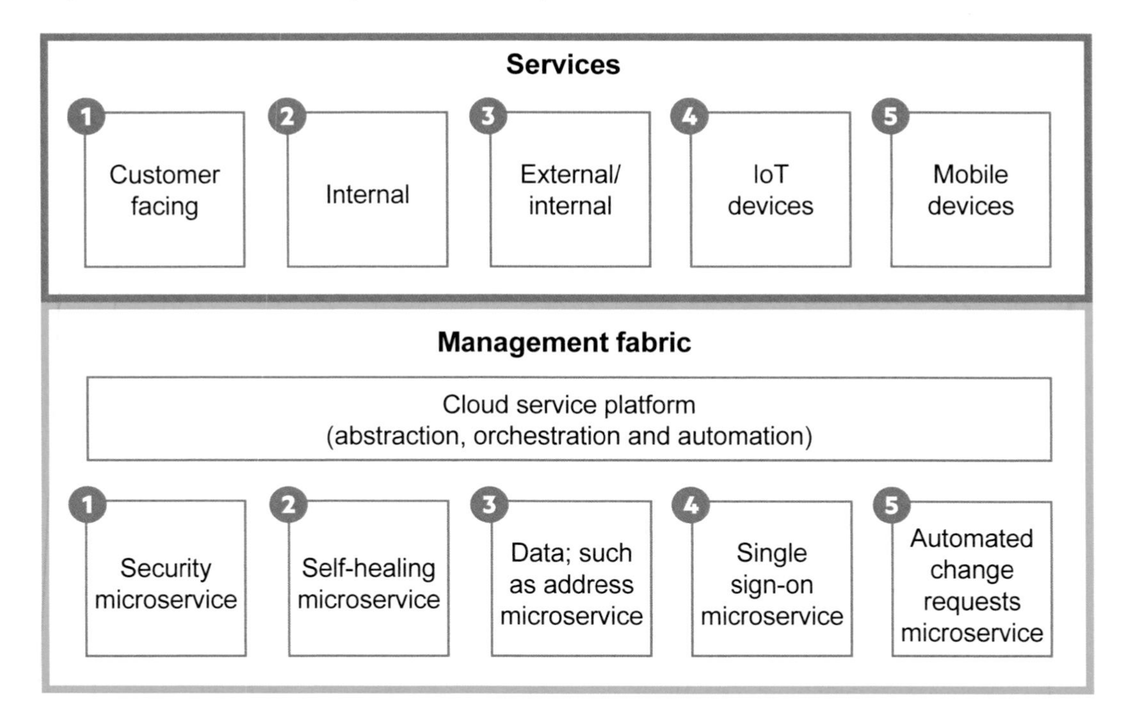

Solutions and Services

The solutions and services that run on top of the platform drive the design of the microservices, orchestration links and automation rules required for the platform as well as the placement of workloads within the platform. They also affect the communication strategy, security and frequency of changes. The cloud service delivery team should consider the following categories of solutions/services when building the cloud service platform:

- **Customer-facing services** are consumed directly by the company's customers, so they usually require more communication with stakeholders, more time educating users and more stringent SLAs.
- **Internal services** include solutions and services that employees use to take care of customers and run the business. Examples include accounting, order entry, inventory and call center.
- **External/internal services** are not customer facing but they do affect user satisfaction and costs. CRM systems, self-service portals and unified messaging platforms are examples.

- **Internet of Things (IoT) devices** connect to the IT infrastructure and can affect costs, compliance and security. Because IoT is new, companies often overlook this category until a security breach occurs or there is some other compelling reason to address it. IoT produces a lot of data, so other areas to consider are storage, processing and impact on big data analytics. This category requires a lot of thought and consideration regarding how to incorporate it yet separate it to reduce security risks.
- **Mobile devices** pose additional security and compliance risks. Whether they are treated as a separate category depends on corporate policies for bring your own device.

Management Fabric

The management fabric enables the solutions and services running on top of the platform to access the microservices and functionality that reside in the foundation. It provides three key functions:

- It serves as an **abstraction layer** that hides the implementation and functional details of microservices from the solutions and services. It ensures that the appropriate rules and policies are applied to each solution and service. The security microservice rules that apply to an IoT solution are likely to differ from the rules that apply to a CRM solution. The abstraction layer ensures that each solution is linked to the appropriate rules and policies.
- **Orchestration** contains policies and actions that apply to solutions/services, data, microservices, configurations, networking and users. This functionality enables the composition of services based on criteria such as the lifecycle of the service, data use (test versus production), location and/or access enabling the composition and decomposition of a service on demand. The policies cover everything from control over who can access a solution, service or microservice to the data that belongs to each solution and service.
- **Automation** includes special tools or interfaces that automate the orchestration links between the microservices and the solutions/services that access them.

Microservices

In Phase 2, you identified common microservices that must reside in the platform to support multiple solutions and services. As Figure 23 shows, common microservices include security, self-healing, single sign-on and so forth. This layer also includes such microservices as eSignature, PDF readers and other functionality that is shared to avoid the cost of building these microservices into multiple solutions/services.

Build and Test the Cloud Service Platform

Next the cloud service delivery team builds the platform described in the platform BRD. The team identifies required adjustments and clearly defines what the platform will and will not do for each release. If issues arise, the team prioritizes them, addresses the most pressing needs first and schedules less-critical ones to be addressed in later releases.

Step 2: Assess, Address and Automate

Under the direction of the cloud service delivery manager, cloud architects and solution/service owners conduct a deep-dive assessment of pilot candidates, looking at everything from footprint to hardware dependencies.

Very rarely is moving a workload from one environment to another a one-for-one match because efficiencies in technologies, different blends of technologies and enhancements in the environment may have a significant impact in footprint requirements, resource skills, location of dependent service components and users.

In this step, you assess and address the requirements of the services to identify the appropriate cloud environment over the life of the service. Both business and IT people assess the pilot candidates and decide which three to five solutions/services are the best ones for the pilot. This assessment often uncovers additional costs, tasks and requirements. These discoveries help the CCoE determine which solutions/services make the cut for the first release of the cloud service platform and which ones are postponed.

The CCoE may decide to push a solution or service out to a later platform release because the amount of work required may affect the overall platform release or other pilot solutions/services. Or the CCoE may decide to release a subset of a service on the platform with a smaller set of functionality integrating to legacy solutions until the full service is ready. (Not every component of a solution/service has to be included in the first iteration.)

Parallel Activities

The cloud service delivery manager manages a subset of operations people who review, categorize and recommend services for subsequent phases—in much the same way a chef manages prep cooks in a kitchen.

Automate Enforcement Rules, Policies and Guardrails

The rules, policies and guardrails created in Phase 1, Steps 4 and 5 need to be automated to ensure enforcement with the pilot and subsequent services that run on the cloud service platform. Figure 24 shows the cloud positioning system created in Phase 1.

Where possible, the cloud service delivery team creates microservices that automate rules and policies via the management fabric. The first platform release may require manual application of some policies, rules and guardrails until all automated microservices have been built. For the assessment, pilot solutions and services need to be able to fully exercise and test the automations for each guardrail, policy and rule. Testing those guardrails is an essential part of the build and test phase prior to releasing the service to production.

Figure 24. **Cloud positioning system**

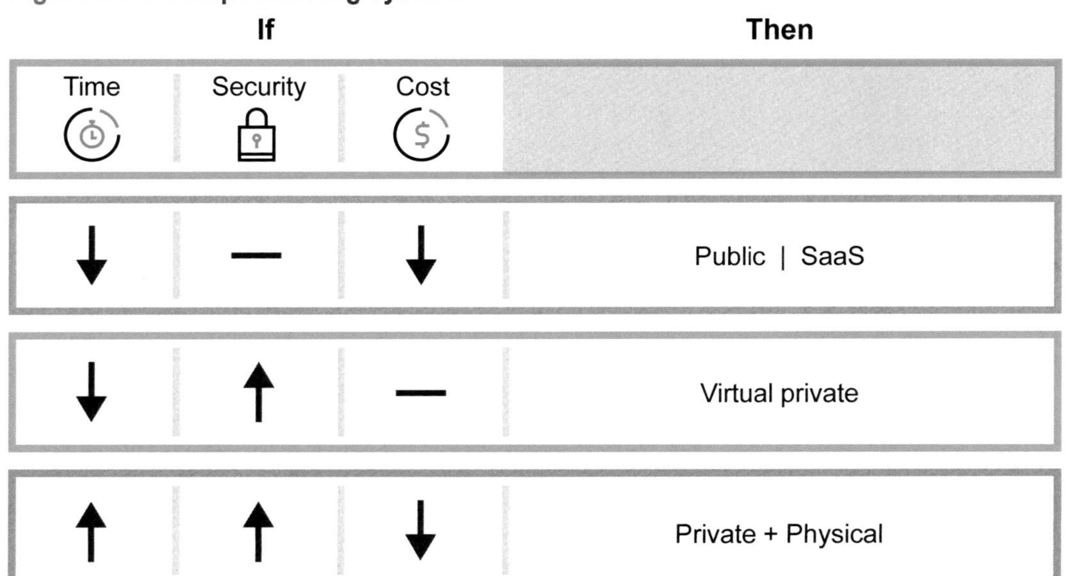

Reduce Services Footprint

In Phase 1, the CCoE requested cloud architects and competency centers to collect data regarding the services being considered for the initial offering of the cloud service platform. The cloud architects use the information provided by the competency centers to:

- Analyze resource consumption of existing on-premise systems to determine if their footprints can be trimmed and optimized to operate more efficiently in the cloud.
- Determine the footprint in the cloud based on the optimization of hardware, software, storage and other resources that will be achieved when a solution/service moves to the cloud.

During this step the team notifies solution/service owners of potential footprint reductions, identifies any factors that might affect footprint and forecasts usage for solutions and services that will run on the cloud service platform.

A large retailer began migrating its service portfolio to a private/virtual private cloud infrastructure. Instead of optimizing to reduce the footprint, IT used the same number of virtual and physical machines in the cloud that were used prior to the move. A careful analysis after the move showed that many services were not fully utilizing the VMs procured for the cloud infrastructure. Some services required only 12 percent of their original footprint because of the higher efficiency of the new hardware implemented for the cloud. Approximately 7 million dollars' worth of VMs that had been purchased the previous quarter were decommissioned because they weren't needed.

Stay, Go or Pull the Plug

The cloud service delivery manager, cloud architects and solution/service owners categorize each of the pilot solutions/services so they can test their theories and assumptions, balance the portfolio rollout and create a repeatable process for onboarding future services. The team must make tough choices regarding whether to migrate part or all of each solution or service, leave it on the legacy infrastructure or eliminate it and replace it with a new one. Use the following categories to facilitate the decision process:

- **Low effort** – Services in this category don't have major security, compliance or cost (technology or otherwise) implications or other restrictions preventing their migration to cloud.
- **Medium effort** – These services can move to the cloud but require some enhancements around technology, security, compliance or other factors as indicated by the cloud service delivery team.
- **High effort** – These services must be re-evaluated because they require significant enhancements, are costly to move or involve legal restrictions. A solution/service in this category may be moved to the cloud, but the team needs to pay close attention to restrictions.
- **Not migrated** – These services will not be migrated because of legal restrictions, low user count, low service usage or intent to retire or replace the service with another tool.

This categorization enables the CCoE to queue up solutions and services for the move to cloud. Low-effort services move to the front of the line for the first wave of migrations. Priorities for medium- and high-effort services can be set based on contribution to corporate goals and such factors as security risks, cost and ability to facilitate migration with technologies such as containers, application virtualization or application transformation technologies.

Vendor Negotiations

Bringing solutions/services onto the cloud service platform often involves negotiating with vendors for new or modified contract terms. Specialists from procurement and/or contract management should participate in this effort. These people initiate communication and negotiate contracts as needed. They should work closely with and under the guidance of the CCoE to prioritize the list of approved vendors.

Current contracts were likely negotiated in silos. Renegotiation offers opportunities to save money through volume discounts and to obtain more favorable SLAs. A creative negotiating tactic is to include a clause stating that the subscription period doesn't start until the product is rolled out to a specified number of users that is higher than the number of users in the pilot. Another way to avoid paying for licenses that won't be used until the pilot is completed is paying consumption-based subscription fees during the pilot.

Step 3: Align and Refine Plans

In this step, you refine and finalize the business case, the plan for timelines and resources, and the platform BRD. During Phases 1 through 3, you completed initial planning documentation from executive-level business plans to technical requirements from the architects. As the BRD, implementation and business plans have morphed with new requirements and inputs, they may have been knocked out of alignment from each other. The purpose of this step is to bring them into alignment, ensure the platform is capable of supporting the pilot solutions and services and to obtain executive approval on the final selection of pilot solutions and services.

The cloud service delivery manager evaluates the documents and identifies any anomalies or misalignment across the BRD, implementation plans and business plans. The cloud service delivery manager identifies gaps or discrepancies, updates the documents and then seeks approval from the CCoE for approval of the realignment. This alignment ensures that everyone from the people who created/approved the original business cases, the service delivery leaders and program managers understand what to expect and that new budgets are procured and changes understood.

This step is critical not only to reflect the current and target state for the first release but also for historical reporting and tracking success metrics. These updated documents are essential for the final review and creation of the final rollout plan. The architect, cloud service delivery manager and the solution/service owners are key players in this effort. They have done much of the groundwork in earlier phases, so this step is one of sharpening the pencil point for accuracy, feasibility and risk mitigation.

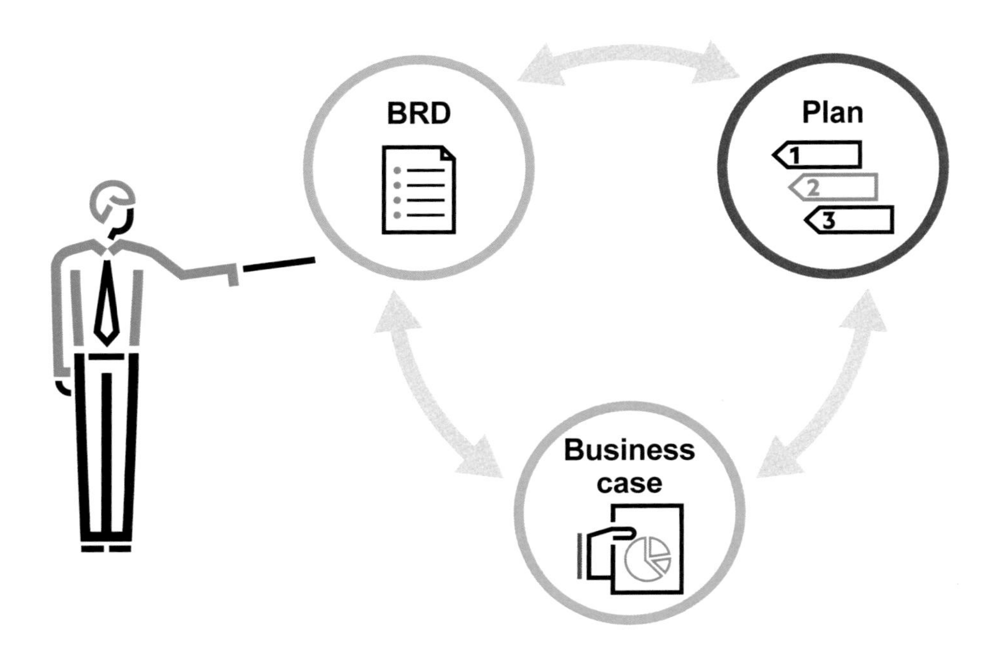

Assumptions and Risks

Addressing the assumptions and risks identified by the CCoE in Phase 1 Steps 4 and 5 is a vital part of integration and iteration. This includes reviewing heat maps for the service, resources and platform. There may be resource constraints around such areas as integration or setting up a lab environment. If so, the CCoE needs a plan to verify that the assumptions are correct.

Other areas to consider include the availability of capabilities from technology and cloud providers. One company interviewed for this book reported significant delays in a pilot because no one on the CCoE realized that the APIs required to connect two SaaS solutions were not yet available, causing a project delay of nearly six weeks. When the APIs did become available, the company was the first to implement them. That put the cloud team in the position of "testing" the APIs, which caused additional delays. Cutover and contingency plans should be created at this point to address risks highlighted by the team.

Timelines and Iterations

Timelines vary depending on the scope of the pilots and where the company is with respect to cloud maturity. A company that has already implemented a private cloud and is now gaining control over public cloud sprawl can probably move to hybrid cloud more quickly than a company that hasn't implemented a private cloud. The latter company will need to allow additional time for planning as well as tuning, integration and timing to mitigate risks.

Pilot plans should call out any blackout periods during which infrastructure changes affecting the pilot solutions/services might impact timelines. For example, retailers typically have a blackout period starting just prior to Black Friday and continuing through the Christmas holidays, and universities have a blackout period during registration at the beginning of each term.

Pilot plans should call out any blackout periods during which infrastructure changes affecting the pilot solutions/services might impact timelines.

Timing should take into account services, versions, resources and users. For example, if a company is deploying an integration to a SaaS solution but the contractor who is providing architectural services is on vacation, the CCoE needs to adjust the timing of the integration accordingly.

Figure 25 illustrates that service lanes for resource utilization have to be calculated and measured carefully to avoid hot spots (resource constraints) or cold spots (underutilized resources). It is also critical for determining overall time to value for a given service based on workload in the critical path employee queues.

Figure 25. **Resource velocity**

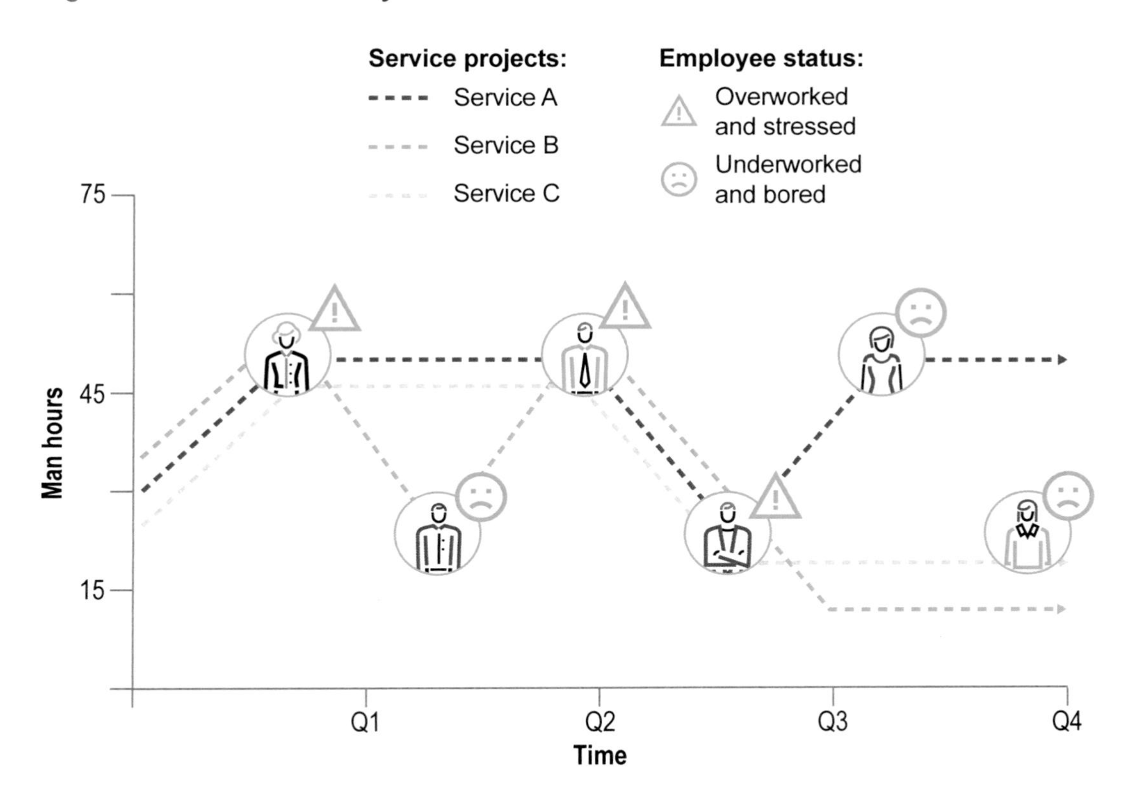

After you have determined optimal resource utilization, timing and overall costs for each solution/service, consolidate the costs and considerations for presentation to senior management. The goal of this activity is to provide a high-level overview for the CCoE, stakeholders and C-level executives and obtain approval on the final plan. The presentation covers final costs, requests for additional funding as required, anticipated timelines and any risks that could significantly affect the company, a specific solution/service or a team.

Figure 26 shows a typical diminishing returns graph that demonstrates to executives that delays in cloud implementations or use of specific infrastructure will cause a tipping point to be reached where the costs exceed the benefits. The light blue line illustrates the diminishing returns versus the project returns on the purple line. Graphics like these help identify tipping points for hosting workloads on third-party cloud in comparison with refactoring services in the stack for the private cloud.[17]

---

17. *iSpeak Cloud: Crossing the Cloud Chasm* contains other sample tools and calculations to help in examining costs.

Figure 26. **Resource velocity**

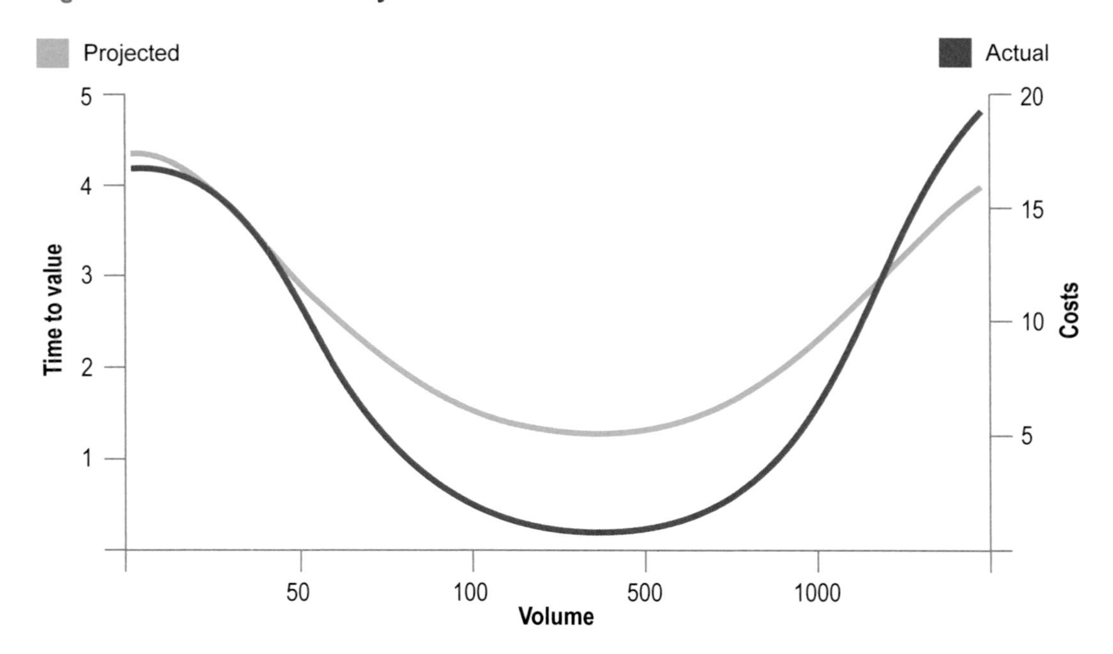

Step 4: Develop Communication Vehicles

The primary focus of this step is to create materials that communicate progress, plans and successes to executives, peers and end users on an ongoing basis. Consider assigning a technical writer or editor or hiring a communication specialist to create and execute the communication plan.

Regular communications regarding what to expect within what timeframes has been provided and all communication vehicles are functional. Communication plans should address executive, peer and user communication with the appropriate level of detail and tools to support transparency.

Executive Communication and Dashboards

Until this point, the team has performed a number of analyses and costing exercises. It's now time for the CCoE to review with executive leadership the results of these efforts. This presentation serves as a checkpoint for assessing progress, evaluating the completed work and obtaining approvals and guidance prior to conducting the pilots. Items for communicating to the executive team include:

- A **heat map** showing the potential risks based on the top pilot candidates. Sources of risk include the impact of replacing legacy systems, using a new third-party cloud provider or using unproven technologies or processes.
- **Summary of KPIs, projected benchmarks and reporting** to apprise management of what to expect during the pilot, rolling out the solutions/services beyond the pilot and other issues executives need to be aware of. The summary is intended to be an executive-level review. The communication should indicate how frequently updates on the pilot will be issued, any tools or dashboards that will be available for viewing the updates and who will have access to this information.
- **Tool gap analysis and review** involves creating a general list of the tools you'll need and examining how current tools will work with the cloud service platform. For example, the cloud service delivery team checks to see if the current monitoring solution can operate in a hybrid cloud environment. The CCoE reviews the tool gap analysis prior to tool selection or enhancement. The results of the analysis and review may affect the selection of pilot solutions/services. If the team learns that the hybrid monitoring solution does not function with a particular solution/service, the CCoE may recommend keeping the solution/service in the legacy infrastructure until it can be monitored in the cloud.
- **The cloud journey infographic** illustrates successes to date, targeted release dates and versions to set expectations. The infographic shows how many solutions/services and users will be brought onboard to the new platform and in what timeframe. (See Figure 27 below.)

Figure 27. **Rollout chart**

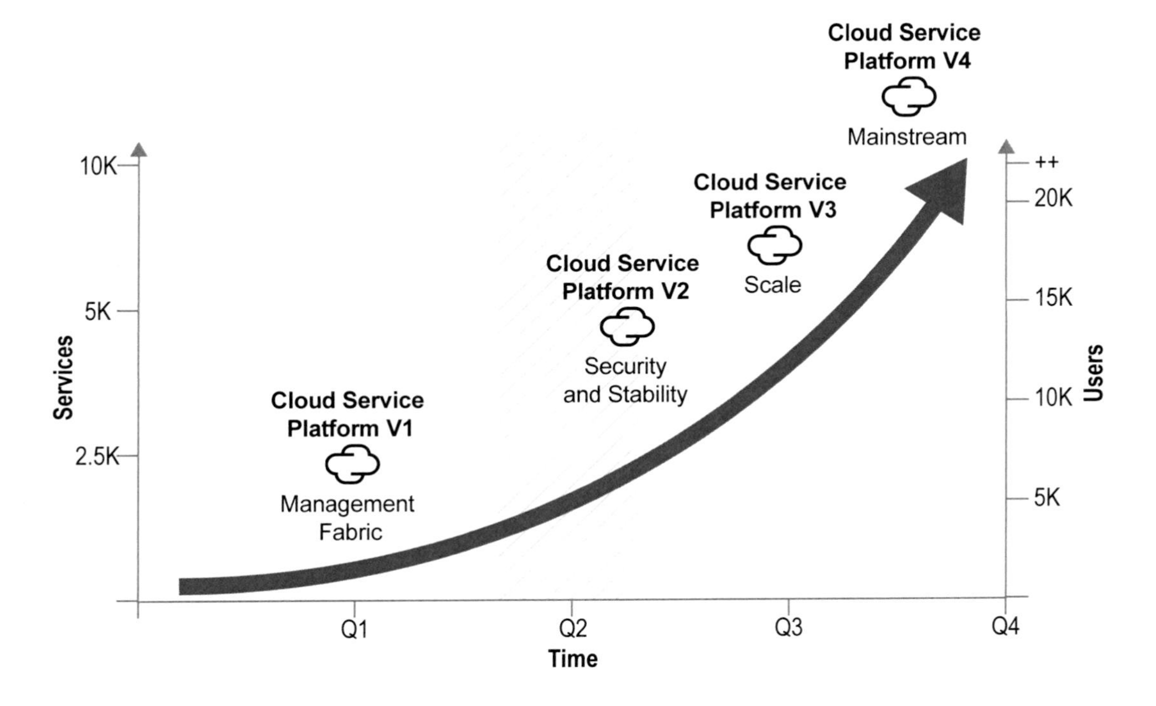

Peer Communication and Dashboards

The cloud service delivery manager needs to establish a regular cadence of communication, not only into the CCoE but also out to owners. Creative ways to communicate information include:

- **Road shows and webinars** to review timelines, KPIs, expectations from each of the players and next steps work well for larger organizations.
- **Wikis, town halls and conference calls** run by the cloud service delivery manager enable smaller organizations to take a more intimate approach to building stakeholder buy-in.
- **Requirements and process enhancements documentation to follow compliance by design mandate.** The process for designing in security, privacy and compliance is published and training has been scheduled for service owners to understand what they must do, why they must do it and how they connect into the cloud service platform and supporting environment.

> "Compliance by design is critical to overall success in the digital era."
>
> — Marios Daminades
> Partner E&Y, Board Advisor ISACA

- **Change management process automation requirements** have been approved for building out an automated system.
- **Enhancements to the traditional incident and problem management process** have been clearly defined, communicated to the support team and approved for automation by executive leadership.
- **Cutover and contingency plans from Step 3 have been documented** and conducted and the results shared. Each solution/service must be tested for the rollout of the first service and integrations, followed by a slow rollout of additional services and integrations. All contingency plans and risk analyses should have been conducted, reported and signed off by executive leadership.

The cloud service delivery manager should also communicate progress to a broader audience to generate enthusiasm and let people know what to expect and when.

End-user Communications

End-user communications should be part of the overall planning process. Avoid issues by working closely with the executive program management office, department leaders and user experience leaders to communicate changes before they are rolled out. One interviewee reported that the service desk was flooded with calls because of an icon color change that end users mistook for a virus.

Unless a change directly impacts the user experience, the CCoE should handle communication similar to the way Netflix™ and cell phone providers announce infrastructure changes. They state that higher-level system enhancements will be implemented, they roll out the change and they notify users only if there is an error with the change outside the initial rollout window and rolling the change back doesn't address the issue.

Create Supporting Materials

Part of cutting through the cloud clutter is clearly articulating the vision, use cases and best practices for the cloud service platform. Develop the following materials to help the cloud service delivery team achieve that goal:

- **New policies or exceptions to existing policies** – If you are automating something, you need to document and communicate that change to stakeholders. For example, if you're automating the change review process, be sure the service desk and change advisory board are aware of the new process and their role in it.
- **Documentation covering APIs and integration points** – Use wikis, electronic newsletters, knowledge bases, printed materials and other formats to provide solution/service teams with knowledge that helps them take full advantage of the platform.
- **Known issues or defects for each microservice** – Document known defects along with workarounds and expected remediation. This information reduces service desk calls and escalations.
- **Business capabilities, supporting features and highlights** – Create a platform features, functions and benefits document. Infographics showing business capabilities and cost savings can make this information easier to understand.
- **Frequently asked questions** – Write a question-and-answer document that presents information in an easy-to-understand format.
- **Release schedule** – Publish a cutover schedule laying out timelines, number of users, risks and any anticipated service disruptions.

Step 5: Review Final Checklist for Pilots

In this step, the CCoE does a checklist review to ensure that the top three to five solutions and services identified for piloting are the right choices. Here are some sample checklist questions.

Do the services have the right number of integrations?

- How many integrations does each solution/service require? If more than five integrations are required for any solution/service, can the solution/service be scheduled for a later release of the platform?
- Do any solutions/services need to be scheduled together?
- Do the solutions/services fully exercise the microservices scheduled for the first release of the platform? If not, should you pull additional integrations into the first release or push microservices out to a later release?

Test Critical Requirements for Scale

The platform BRD identifies the minimum set of features the platform must have to test for scale. These include security, compliance, number of users, automation and integration. The cloud service delivery platform team needs to run scalability tests against each microservice in the platform and answer such questions as:

- Do the platform and microservices meet minimum requirements for the first release?
- Did load test simulations for the maximum number of users, systems and data run successfully? If not, should specific microservices be deferred to a later release?
- Does the platform meet the minimum security requirements for the pilot solutions/services?
- Are there workarounds or alternatives that can be used in the interim?
- Does early testing prove that the platform meets minimum requirements for availability, supportability, reliability and serviceability?

Test Communication and Governance Processes

- Have you conducted a simulation of the entire communication and governance process? What were the results?
- Have the communication plans and vehicles been tested?
- Do all email aliases, wikis, hover links, input forms, websites and other programmatic solutions work?
- Are the new governance and communication processes well documented?
- Is training available for the new governance and communication processes?
- Have process automations and exceptions to current processes been documented and communicated to users and stakeholders?

Costs, Compliance and Timelines

- Do the CCoE and stakeholders clearly understand costs, compliance concerns and timelines for the pilot?
- Will the pilot timelines address new regulatory requirements in a timely fashion when the solution/service scales? If not, is an alternative needed?
- Is adequate funding in place for the pilots? If not, how much more is needed and what is the plan to procure it?
- Have all key stakeholders signed off on acceptance of risks and the specifics of what the first release of the platform will include?
- Is it clear who will deliver what and when?
- Do the owners of the pilot solutions/services have resources assigned and are they ready to execute during the pilot period?
- Have the solution/service teams dedicated time to provide feedback and work though any rollout issues?

Test and Measure Chargeback/Showback

- Can the platform measure and report on KPIs and provide chargeback and showback data?
- Is it clear what units are being measured and how much they cost?
- Can the stakeholders get detailed reports about chargeback or showback from dashboards?
- Are there other KPIs that must be measured from day 0 such as network latency, user impact and minutes of downtime.

Now that you've completed the steps in Phase 4, you're ready to move ahead to the execution and evaluation phase.

Figure 28. **Summary of Phase 4—Calibrate vision to reality**

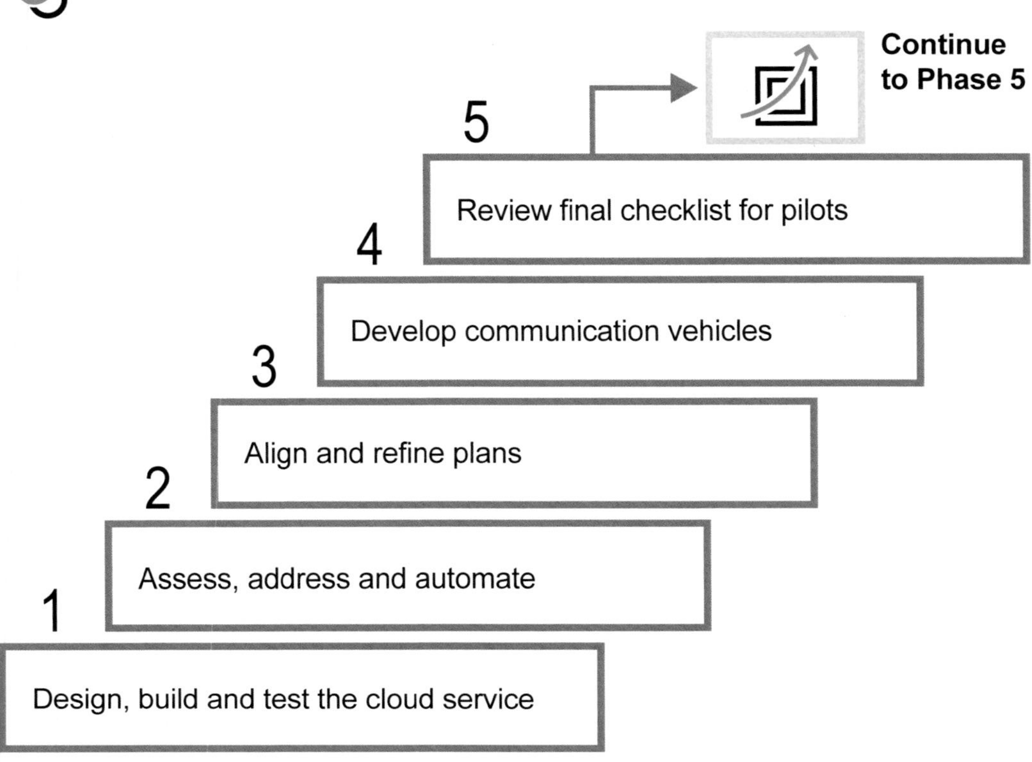

Words of Wisdom from the Trenches

Pat Spica
Managing Director
IT Service Management

Picking Your Pilot and Automating ITSM

Disrupt the Execution Arm

For our cloud initiative, leadership hand-picked people, removed them from their existing roles and gave them a completely new set of rules. In 18 months, the company was way ahead of industry peers with respect to cloud and the momentum propelled our cloud strategy forward. The lesson learned: Don't use your existing team and execution mode for cloud. Hire a disruptive manager to manage it.

Don't Just Move: Clean House First

Creating a CCoE is about adopting a new environment. Part of the adoption is letting go of manual processes and determining what to bring forward and leave behind. Look for areas that are ripe for automation and disruption such as change, incident and problem management.

CCoE members should understand how existing technologies, users and processes will need to be adjusted to the new environment. Recording changes as they occur is critical for long-term maintenance and understanding. That does not necessarily mean that the manual processes to do so cannot be automated to reduce the time spent reviewing and understanding the change. The same can be said for incident and problem management.

Uncover All Costs

Costing is about making sure you know what the real price tag will be. The goal cannot be implementing cloud or hybrid cloud. It has to be a specific business goal such as reducing the footprint of the data center or enhancing customer experience to increase retention.

Learn from Mistakes

Analyze successes and failures and use governance and understanding to move things forward. Put a governance framework in place so if something doesn't go well you can learn from mistakes and not repeat them.

Sharpen Negotiating Skills

Having shrewd negotiation skills and the ability to understand the right sets of cost controls is a huge advantage when dealing with cloud providers.

PHASE 5

EXECUTE AND EVALUATE

"The cloud is the pivotal intersection where users, business and technology requirements come together. Although there are many tools and technologies, it is not just about what you can technically do. Success depends on how quickly the company can absorb, support and adjust to the change the cloud brings."

—Lorenzo Hines
Executive Consultant at SVP Level

Phase 5 Objectives

Execute and evaluate portfolio plans approved by the CCoE. Evaluate success based on the specified KPIs to feed into continuous improvement.

Issues and Clues

The following table lists issues the CCoE addresses in Phase 5 and provides actual quotes from IT leaders interviewed for this book.

Issues	Clues
Communication gaps	"As usual, our biggest challenge has been dealing with people issues. Coordinating across groups during rollout is cumbersome and sometimes chaotic. More often than not we fall into stall from vendors or internal teams or inertia from both." –IT Director, QA
Inefficient processes and legacy tools cause delays and increase costs	"It would take 15 minutes to spin up the VM but two weeks to get access permission and another month for the database. And leadership wonders why the business is working around IT." –IT Director "The change process is archaic. We email change requests to 15 people. The change advisory board sends the documents back to provide approval for deployment. Multiply that by 600 requests a week and you can see why we're experiencing higher network and data traffic." –Director, Cloud Service Delivery
Monolithic implementations equal monolithic failures	"A lesson Healthcare.gov taught us is there is no such thing as too big to fail. Traditional monolithic projects are risky because the market changes before they see the light of day." –CIO

Issues	Clues
Unrealistic and/or unenforceable policies	"If there is shadow IT going on, something is wrong. Big bad corporate forcing things down users throats won't work. People will do what they have to in order to do their jobs." –Business Leader
No thought given to global impact	"Cowboy teams within IT come up with their own processes for cloud. They select tools that don't integrate with current solutions and expect everyone else to conform." –VP of Infrastructure and Ops "It's hard to keep top talent. Our people are stretched too thin by too many tools that essentially do the same thing. We have seen nearly 75 percent turnover in operations and support in the last year. This makes it harder for those who are left." –Enterprise Architect
Dev and ops teams cannot exit the pilot	"Day 0 support was nonexistent. The pilot was released in 18 days but it took months for tier-3 engineers and architects to extract themselves from day-to-day implementation issues." –Business Leader "The pilot went really well, but scaling after the pilot didn't. Once we hit 1,000 users we started to see communication, performance and scale issues—across people, process and technology." –Program Manager

Step 1: Lay the Foundation for the Pilot

The first step is ensuring the right communication and structure is in place among the three execution areas: the business, engineering and operations. Collaboration and communication across these groups ensure that everyone knows what needs to be done to achieve success. One representative from each area acts as the execution focal point and reports into the CCoE. The tasks for each group are:

- **The business** – The owners of pilot solutions and services appoint one business leader to take responsibility for the pilot and manage the interactions with business leaders and end users.
- **Engineering** – The cloud service delivery manager or an appointee from the cloud service delivery team serves as the point of contact for communicating with technical resources. This person communicates across to development and operations and down to the competency centers during the pilot to ensure that technical requirements are met, technical debt is minimized and resources are protected from thrash.

- **Operations** – The cloud service delivery manager or an operations-focused person from the CCoE serves as a focal point for the operations, support and governance teams. This person handles communication and coordination across the organizational change, development, program management, service desk, network operations center, client and infrastructure teams.

Ensure success by focusing on people

- Increase staffing for development and operations so you have sufficient resources to support the pilot.
- Brief support team members on the initial process for requesting enhancements and new features.
- Map the progression from legacy roles to new roles as services and tools are migrated or replaced.
- Eliminate FUD
    - Make sure all resources have a clear vision, focus and pathway for success.
    - Manage expectations and incoming requests while pushing the team to maintain focus on execution and quality.

Day 0: Approval of Technology and Launch Prerequisites

Laying the foundation for the pilot includes getting signoff and participation across affected groups to release the solutions/services and platform into production prior to launch. Be sure to obtain this from the primary technology providers such as the cloud service delivery team as well as teams responsible for supporting or dependent solutions and services. Ensure that you have the right approvals from every affected area: ITSM, operations, quality control, and tools and technologies.

Each area must have plans in place, resources assigned and automations executed and tested as indicated below:

ITSM

- Architectural, security and audit approvals must be signed off and any change advisory board documentation and requests must be complete. If one of the microservices automates the change process, the automation sequence and signoff have to be completed prior to service rollout.
- Asset inventory controls and mechanisms need to be in place to ensure that assets in third-party clouds and on premise are licensed and monitored. This can be accomplished with a microservice that checks the contracts database for the appropriate license type before allowing replication or cloning of a VM with a service or application component.
- Incident/problem management process automations are baked into the incident and problem management process to accommodate self healing, rollback, roll forward and other configuration tasks defined in the microservices BRD. For example, if a service is nonresponsive, a monitoring microservice can roll it back or forward and open an incident. If the automation fixes the issue, the incident is closed. If not, it is escalated to the next support level.

Operations

- Monitoring across hybrid clouds for every provider in the cloud service platform are in place. Monitors should determine not only if a solution, service or microservice is up or down but also if it is actually working.
- Tools to collect and analyze KPIs (IT operations analytics) for executive briefings are ready to capture all metrics identified in the business cases and BRDs. The more automated these tools are the better. For example, if there's a requirement to pull data from ERP, project management and defect tracking systems and create custom logging to measure performance, build a microservice that performs those steps.
- Backup system access and control for business continuity are ready to go. This includes the ability to quickly catalog, snapshot and restore a service based on a dependency map across any part of the hybrid cloud environment. If the private cloud cannot serve up a service due to technical issues, the service can be restored on a virtual private cloud from a backup and connected to elements stored on other clouds.
- Mechanisms for feedback are in place and automated from user through development, including email, forms and other means of communication identified by the CCoE. For example, it's possible to create forms with cron jobs such that when a change ticket is updated so is the repository of errors reported by monitoring and log analyzer tools.

Quality Control

- Load testing of the platform and microservices has been completed for the first release and, if possible, the initial three integrations. Load test the services on top of the platform along with the critical parts of the system such as network and integration links, particularly to converged communication or multichannel solutions. If the project does not allow for the cost of broader load testing, focus testing on potential breakpoints.

Tools and Technology

- Service environment management tools and process have been implemented. Because of the variety of environmental variables across a hybrid cloud stack, you will likely need additional tools for releasing services to production, network access control, discovery, orchestration and other critical configuration and operations tasks. Tools and processes are required to manage every aspect of the service stack to enable full mapping, especially of the data, type of data, location and access as part of the service. Interviewees for a this book indicated that of all service stack components, data is perhaps the most critical component to manage, yet it is the one for which the fewest number of comprehensive tools are available. Integrated tools that manage composable service components across the stack and hybrid environments are an emerging area. It is important to look at the existing tools migration pathway and new tools to determine the best route to value for your requirements.
- Configuration management tools for container/VM/application management are in place. You've identified not only the VM or container but also how it relates to the entire service stack. Such tools should be able to snapshot, clone, map and recreate all aspects of the service stack to enable portability across hybrid environments. For example, a company deploying containers may have adopted a container management tool to deploy the entire container or VM cluster of the solution/service across hybrid clouds as needed—for example, moving from a private cloud to a third-party cloud.
- Service and/or application maps and dependencies are complete. Maps detailing dependencies across networking, storage, data, containers, microservices and configurations should be available. Each one should have heat and dependency maps. Each hotspot and dependency should be monitored and correlated across hybrid environments in an automated way to reduce risks.
- The service catalog helps control access by ensuring that only people who are part of a pilot user group are able to view and select pilot solutions and services. Over time, the solutions and services will be released to a larger and larger set of users. The number of users and access controls should be defined by solution/service owners and the organizational change development representative on the CCoE.
- The self-service portal across cloud providers allows people to provision the slice of compute and/or microservices needed for the services they are creating. Once a request is submitted, the portal automates the provisioning of the requested components, including data, storage, networking, security and access control. Clusters are programmatically created and reviewed based on user access control permissions and signoff from appropriate budgetary owners.
- Security policies, provisions and actions are in place and programmatically enforced. Every solution/service in the pilot that accesses the platform is classified as high,

medium and low security or exception (in which case a manual override by two employees and one executive is required for emergencies). The cloud service delivery team creates and deploys microservices to check pilot solutions/services for compliance prior to day 0.

- Resource pools, chargeback/showback and definitions have been programmed. Cloud resource pools and definitions for slice of compute, storage and access have been programmed for access from the self-service portal for provisioning on demand.
- Policies and infrastructure have been automated. Policies are in place and defined through an automated orchestration system within the management fabric to enforce workload placement based on data, security, software license, user access controls and other attributes to obtain proper approvals and avoid nomadic workloads. For example, one large financial institution implemented a solution to check license access from a microservice prior to allowing a VM or container to be cloned or migrated to a third-party cloud for the 20 percent of applications that require 80 percent of ongoing maintenance. These applications are called the "money" applications because they have the biggest risks and costs if there are issues during an audit.

Super User Swat Team

Involving superusers from the start of a pilot helps reduce shadow IT. Initial pilot groups should include the change leaders/agents from customer-facing roles for external services and/or actual customers for internal applications. Select individuals who would typically create or find the shadow IT solutions and rally the troops around them.

Superusers help drive pilot success because they are problem solvers with a vested interest in making not only their jobs but the jobs of the people around them better and easier. If superusers believe in the cause and have input into the process, they are more likely to become champions who are committed to its success.

Create a superuser swat team to provide feedback during early development stages and the pilot, support for their peers and ongoing innovations using the platform. Key tasks for a superuser swat team include:

- Capturing and communicating end-user feedback on flow and performance of the solution/service along with requests for additional integrations, services and functions
- Helping peers understand and comply with policies created in Steps 4 and 5 of Phase 1
- Reporting KPIs with examples of intrinsic benefits that can be incorporated into executive leadership reports and infographics

- Participating in weekly feedback sessions with the cloud service delivery manager and sponsoring executives from the business

Even if superusers don't participate in all the activities listed above, make sure they are involved in the initial pilot. The closer the end user is to the development of the process, solution and supporting services, the higher the probability that the solution will hit the mark with respect to usability and user productivity.

Step 2: Implement and Measure

It's now time to roll the cloud service platform for the pilots into production or simulated production, depending on how the pilot has been designed. From day 0, the cloud service delivery team should be ready to support not only the first solution/service to go live, but also all subsequent pilot solutions/services. All processes, prep work and planning up until this point should be evaluated based on the KPIs.

Evaluate Early and Often

The CCoE needs to evaluate the system early and often but without micromanaging the cloud service delivery team. The team needs to be free to make adjustments that facilitate the onboarding of new users and the optimization and deployment of new solutions/ services. Each business, engineering and operations leader should set up meetings to review the KPIs, gain feedback on processes and update the progress of the pilot.

Check the overall system performance at least twice a day. Schedule a daily meeting for each area—business, engineering and operations—to review detailed results. Share the daily results with executive leadership of the CCoE, the CIO and the chief operating officer.

Triage and Treat

At this point, staying on top of incoming requests is essential. The cloud service delivery team must quickly decide which requests to address now, which ones to schedule for a later release and which ones to set aside. Expect a few glitches at the beginning because no code or platform is perfect. You'll need to find workarounds, enhancements and adaptations to make sure each solution/service functions properly. The biggest difference between a traditional pilot and the cloud pilot is the mechanism for rapid feedback, iteration and release.

One company facilitated the feedback/iteration/release process by collocating microservices developers with the business service development team during the pilot. If unclear error messages appeared, APIs didn't work properly or other issues arose, the developers were able to quickly troubleshoot and resolve issues before they became show stoppers.

Harden Platform, Processes and People

Be sure to allocate time to harden not only the technology but the business processes that support it. From a leadership perspective, the CCoE needs to have a measured level of tolerance for people who become frustrated for various reasons. Implement active listening sessions and action plans to adjust people, process or technology. At times, these sessions may turn out to be venting sessions. Let people vent, but then help them refocus on the task at hand and guide them on the path forward.

Test reliability, availability, security, scalability, supportability and serviceability to ensure the platform meets minimal requirements in each area before scaling the solution/service up or out.

During rollout and hardening, enable rapid handling of issues and overlooked requirements by allocating 30% of developer time for unknowns and 25% for hardening.

Plan to Fail after Stabilizing the System

Test failure on a small scale before expanding up and out. Once the platform, processes and users are stable and ready to scale, each solution/service owner works with the cloud service delivery manager to remove or add process pieces to test each component for failure. Testing should include internal and third-party processes, systems and tools.

Scalability is critical, so test reaction times and processes as well as the system. Selectively work with the CCoE and the platform team to determine what you can collectively fail safely, whether it be the support queue, a node in the cluster, a data connection or other component. Be sure to include the security strategist if these tests include security or compliance.

A major entertainment company hired a third party to run security and penetration tests on a new stadium to test services created around a cloud solution to remotely manage lighting and sound. The third-party security firm found vulnerabilities and applied patches before the large-scale rollout, reducing impact of a breach during a game.

Step 3: Calibrate and Control

In this step, the cloud service delivery manager drives an ongoing review of the progress of the pilots to uncover recommendations for changes in the governance framework, cloud platform and solutions/services.

Listen and Observe

Listen to what people are saying and observe how users are experiencing a solution/service before making judgments and identifying changes.

Adjust as Needed

Once the pilot is operational, necessary adjustments and changes are likely to become apparent. The cloud service delivery manager may have to make tough calls, such as pulling an underperforming solution/service from the pilot.

Recheck the Costs

At agreed upon check points, conduct sanity checks to determine if each pilot solution/service still makes sense. Refer to the business case, success criteria and overall justification for the initial microservices and platform to determine the viability of the plan. Ask in-depth questions to help determine if costs and/or controls need to be recalibrated.

Questions to ask include:

- Do the anticipated benefits justify the costs and current state?
- Are any items affecting velocity?
- Can the solutions/services selected meet their SLAs?
- Does the consumption of bandwidth, capacity and other resources match forecasts?
- Are the costs over or under for the pilot?
- Are there upgrades or potential issues with networking hardware, infrastructure or systems required internally to support the pilots and beyond?

A government agency in the UK shaved several million dollars off the pilot by decommissioning and ceasing to pay for underutilized VM licenses and hardware. Procurement also renegotiated a SaaS contract to allocate more time for development and test based on architectural and other enhancements identified during the pilot.

- Were deficiencies or missing capabilities discovered during the pilot? If so, when and how can they be addressed?
- Do the forecasted capacity, SLAs and subscriptions negotiated with third parties still make sense?
- Are there solutions not previously considered around hyperconverged systems or other options that make more sense now?
- Are there areas where costs can be cut and unexpected costs accommodated?
- Are there additional benefits or value in eliminating technical debt not covered in the previous case?

Go through this exercise for the platform and corresponding workloads that have been prioritized to scale additional services onto the platform after all pilots are completed. Select solutions, services and microservices based on overall value to the business for subsequent releases of the cloud service platform.

Iterate and Balance

The CCoE and competency center leaders need to understand workload thresholds and decide when to declare victory to move forward at agreed upon checkpoints in the process. In today's environment, users and companies are not as worried about failing (unless it's a security failure) as they are about recovery. Part of the CCoE's job is to determine next steps in optimizing the platform and the onboarding process for future solutions, services and microservices. The CCoE should review the list of services and solutions identified by the cloud service delivery team and balance the overall plan and portfolio based on lessons learned, velocity of team and recalibration of the plan. Questions to ask include:

- Is each pilot solution/service balanced across solutions to adhere to budgets?
- Will any of the solutions/services hit the tipping point where running the workload on a third-party cloud costs more than on a private cloud?
- Are there solutions/services or dependent workloads on private clouds that are underutilized and, therefore, should be moved to a public cloud?
- Have you discovered nomadic service workloads that need to be contained for compliance or cost reasons?
- Are there solutions/services that need tuning before the pilot continues?
- Are the policies around data, users, workloads and administrative access control sufficient for the current solutions/services? If not, what new ones should be created to continue the pilot or scale for production?
- Do any service workloads need to be collocated for compliance, security, performance or financial reasons? If so, have they been tagged by asset management personnel and segmented based on use?

- Are the service workloads balanced? Should targeted workload pruning take place prior to the next phase of the pilot because workloads are not being utilized or are underutilized?
- Do the solutions/services conform to policies and are they optimized to leverage the platform and microservices?
- Are any disruptive events scheduled for the solution/service or dependent workloads that may affect the success of the pilot? For example, is the service workload in a period of peak usage due to end-of-year audit or seasonal consumption spikes?
- Are customizations to the microservices required to make workloads run more efficiently?

One company had two options for solutions/services to consume the security microservice. During the pilot the team discovered that certain services did not have the right level of permissions to integrate to standard APIs on the network. As an alternative, the development team created a code injection system to insert updates into the service itself.

Step 4: Optimize and Onboard

In Step 4, you optimize and iterate pilot solutions/services to work within the CCoE governance framework and the cloud service platform. You also review the pilot, processes and next steps.

Optimize Platform and Plans

Once the platform is stable and solutions/services are performing at acceptable levels, it's time to move to the next step with precision execution. As the introduction to this phase indicates, Phase 5 focuses on iteration, so you repeat the steps multiple times to continuously improve the platform and the solutions and services that run on top of it. You'll need to ask questions in the following areas before onboarding a large number of additional solutions and services:

Tool Effectiveness

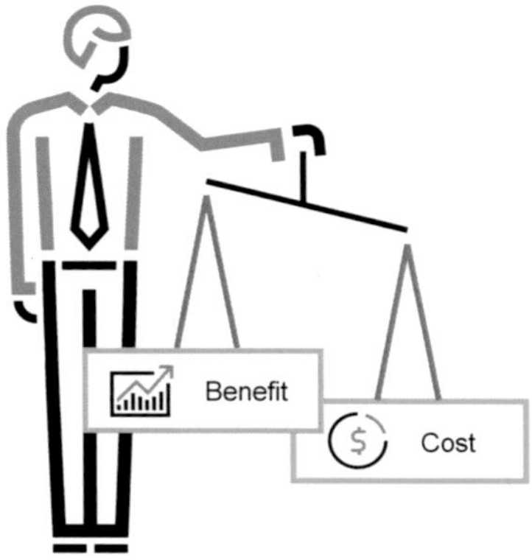

- Are the current tools sufficient?
- Are new tools or technologies needed to address requirements for onboarding additional services?
- Do current and legacy tools work well together?
- Are new APIs or integrations required to automate or check tools to ensure they are not overwriting changes? For example, does the legacy configuration management tool play well with new tools running on the platform?
- Is there a bimodal bridge to workloads or clusters needed to onboard legacy services to the new tools and platform?
- What tools, processes, scripts or other technologies can be consolidated, simplified or decommissioned to reduce friction and complexity?

Cloud Service Provider Consolidation

- Are there too many tools or cloud service providers?
- Which ones should be consolidated?
- Which ones should be kept based on cost calculations and ability to meet SLAs?
- Are there new tools or APIs that were not previously available that will help with migration?
- Does the provider have a mechanism for chargeback/showback?
- What providers were most effective during the support and implementation process?
- Which providers were difficult to work with and should be replaced?

Architecture

- Does the current architecture, including the platform, its microservices and the infrastructure, still meet the requirements for the majority of solutions/services?
- Are tweaks needed to scale, optimize or adjust the plan prior to mass onboarding?
- Have you identified resource constraints to consider across resource workloads?

Policy Optimizations

- Do the current policies and automation tools enforce the desired service and user behaviors?
- Are there issues with solutions that are under- or overutilized?
- Does the company need to implement and automate additional policies to scale?
- Are there mechanisms for eliminating shadow IT, such as programmatically preventing VMs on AWS as a line item expense?
- Do you have reminders/time bombs to snapshot, archive and store VMs that are not utilized or underutilized for 30 days or more?

ITSM Adjustments

- Do adjustments need to be made to ITSM processes? For example, does it make sense to institute an additional level of change approvals for major changes such as data encryption to prevent ransomware attacks?

Asset Management and Discovery

- Are adjustments needed for discovery tools to provide inventory and other controls across hybrid environments?
- Are new mechanisms required for tagging workloads that are lifted, shifted or replicated onto other environments?
- Are there new technologies that can check for license compliance prior to migrating to a third-party cloud?
- Is an injection or update needed into the configuration management database to clarify asset location?
- Are there mechanisms for tagging solutions/services and the assets that comprise them based on service clusters that have been earmarked to migrate to off-premise clouds?
- Is there a mechanism to tag both assets and data workloads associated with the solution/service?

Security and Compliance Optimization

- Is there a mechanism for associating service cluster clones/snapshots with the master and identifying any anomalies in cloning that could be due to social engineering or virus?
- Do machine learning or other security solutions need to be added to the platform prior to onboarding a large number of solutions/services?
- Do additional policies or orchestration optimizations need to be programmed to segment protected assets by location, data type and user?

Communication

- Are the communication mechanisms and channels working?
- Do any areas of communication need to be adjusted? Which ones?
- Do major issues justify revamping parts of the communication process before proceeding to a larger rollout?
- What tweaks and additional information are required as solutions/services roll out to a larger number users?
- Are there roadblocks that must be addressed at the executive level?
- Are there regular meetings to communicate adjustments to onboarding, process and other critical areas?
- What suggestions do participants and users have to improve the process?

Optimization and Implementation

Ask the following questions to keep plans and processes tuned and optimized during the implementation of the platform and solutions/services:

Onboarding Process Assessment

- Does the current onboarding process for the pilot solutions/services flow well?
- Are platform adjustments needed to scale out to hundreds or thousands of solutions/services?
- Does the current rollout plan to onboard solutions/services to the cloud service platform need any adjustments?
- Have the service users, owners and executive sponsors been notified?

Onboarding Selection Criteria Optimization

- Does the CCoE need to adjust how future solutions/services are selected and onboarded?
- Does the CCoE need to adjust how solutions/services are classified from a service category or security risk perspective?
- Should the sizing or estimations of time be adjusted for complexity?
- Do other adjustments need to be made when selecting solutions/services to minimize resource contention or impact to cost or time to value?

Calendar Optimization

- Were the calendar and timelines provided for the list of service clusters targeted for onboarding during the pilot effective? If not, how should these be modified?
- Was the calendar readily available to all stakeholders?
- Were calendar changes communicated in a timely fashion?
- Were emergency changes that affected delivery dates understood and discussed after the fact with the solution/service owners?

Figure 29. **Summary of Phase 5—Execute, evaluate, and iterate**

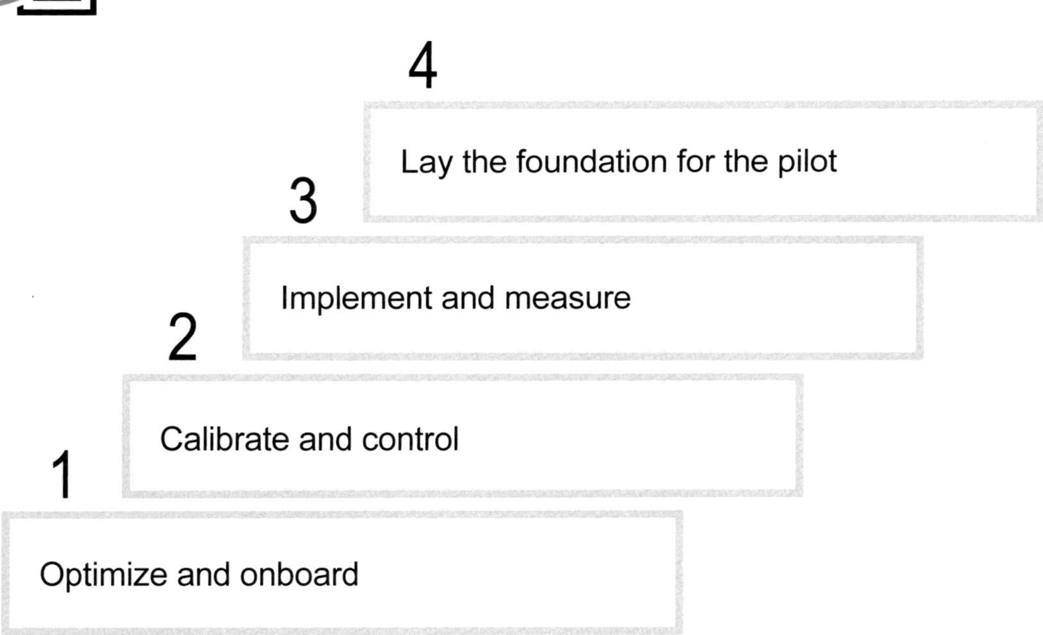

Words of Wisdom from the Trenches

Chris Armstrong
Enterprise Architect and Consultant

From Pilot to Full Production

Extend Communication to Cloud Providers

Collaboration with cloud providers makes a big difference when it comes time to scale to production. Strong relationships enable you to understand the strengths and limitations of each cloud offering. Always have two providers for backup, regardless of how many cloud providers you have for providing services to users. But ensure that you foster a spirit of collaboration versus competition so that if issues arise, providers are willing to work together for the greater good of their mutual customer. The more the cloud service providers are involved in their specific architectures, integration and resource planning for providing solution specifics for your cloud service platform, the more successful you will be in expanding from pilot to full-scale production.

Architect Beyond the Users to IoT and Mobility

The Internet of Things is coming like a freight train down the track. You can plan for it or get derailed by it. Planning involves building competency around the growing number of mobile devices and edge devices/appliances that are connecting to your infrastructure. Be proactive in segmenting and augmenting infrastructure, access and design based on the assumption that these areas will expand quickly. If you aren't careful, you'll end up shutting down shadow IT for solutions and services while shadow IoT and mobility support spin out of control.

Secure the Riches not the Perimeter

Ensure all assets, network, connectivity, firewalls and configurations used to run the system are up to date and well documented. Most security risks are introduced by undocumented/unknown changes. You need to manage and understand new and legacy systems to have a clear picture of risks and architectures. Instead of trying to secure everything all at once, focus on the data and services that matter most. Quarantine data that contains personally identifiable information or intellectual property. Require additional layers of approval for changes to higher-risk services and data. Understand and manage the black space. Managing the data location and staying in compliance is the hardest thing, but you can streamline it with the right precautions and architecture.

GLOSSARY

Term	Definition
Agile Development Model	A group of software development methods based on iterative and incremental development, in which requirements and solutions evolve through collaboration between self-organizing, cross-functional teams. It promotes adaptive planning, evolutionary development and delivery and a time-boxed iterative approach. It encourages rapid and flexible response to change. Its conceptual framework promotes tight iterations throughout the development cycle.
Application Virtualization	Software technology that improves portability, manageability and compatibility of applications by encapsulating them from the underlying operating system on which they are executed.
Business Case	Financial and strategic justification for implementing new services and/or supporting technologies based on a positive return on investment for the company.
Business Requirements Document (BRD)	Document that contains the primary constructs of the requirements for a given IT service. A BRD includes a framework for defining minimum requirements for reliability, availability, serviceability, supportability, scalability and security.
Business Service Management (BSM)	Unification of people, processes and technology across service delivery, service support and service management to proactively address the needs of the business.

CapEx	Short for capital expense. Capital expenses are funds used by a company to acquire or upgrade physical assets such as property, industrial buildings or equipment. This type of outlay is made by companies to maintain or increase the scope of their operations. These expenditures can range from repairing a roof to building a new factory. They can be depreciated over the life of the benefit.
Cloud Calibration	Synchronization and creation of policies and procedures that govern the use of various types of cloud within an organization. It includes but is not limited to policies across organizational units or cloud providers, or cloud bursting or use of different types of clouds based on role.
Cloud Center of Excellence	A cross-functional team that provides leadership, best practices, guidelines and support to enable the implementation of viable cloud projects.
Cloud Governance Framework	Process framework that integrates a cohesive strategy across business groups including but not limited to engineering, business, operations, audit, governance, finance, architecture and product management. All components of the framework tie together through checks and balances across groups to determine best strategy and policy around the use and implementation of cloud computing.
Cloud Service Delivery Manager	Leads the CCoE and cloud service platform creation from ideation to end of life.
Cloud Service Platform	The platform includes a foundation containing microservices and functionality for use by cloud solutions and services plus a management fabric that enables the solutions and services running on top of the platform to access underlying microservices and functionality in a self-service manner.
Competency Center	Established by the CCoE, these teams are tasked with developing cloud requirements for specific functional areas, including customer-facing development, cloud platform, security and compliance, IT service management, finance and operations.
Depreciation	A method of allocating the cost of a tangible asset over its useful life. Businesses depreciate long-term assets for both tax and accounting purposes.

DevOps	A methodology that integrates disciplines of software development and IT operations to create a cohesive, collaborative, integrated process for iterative development and delivery of applications in a secure, predictive and time-efficient manner. DevOps targets quality assurance, delivery, and development teams and processes.
Digital Enterprise	SearchCIO Tech Target defines a digital enterprise as an organization that uses technology as a competitive advantage in its internal and external operations.
Digital by Design	Companies that were born on the internet and if the internet ceased to exist they would be out of business. Technology is their business. Examples include Google, Facebook and Uber.
Digital Native	Person born during or after the introduction of digital technology and who has gained familiarity with digital concepts as a result of interacting with technology from an early age.
End of Life	Decommissioning of older software or hardware that is no longer in use or not used enough to justify the cost to run it.
Epic	The top level of a story map under which individual user stories are mapped.
Hybrid Cloud	An integrated infrastructure environment that leverages both private cloud (on premise or managed) and public cloud (off premise) providers to address capacity requirements for applications (SaaS, PaaS or Iaas).
Infrastructure as a Service (IaaS)	Software technology, also known as cloud infrastructure services, that delivers computer infrastructure. Typically a platform virtualization environment that includes raw (block) storage and networking as a service.
Integration, Tuning and Timing (ITT)	Framework for determining best path forward for integrating multiple workstreams in a complex implementation such as a multilayered cloud strategy based on a number of integrations, amount of tuning and anticipated time period.
Key Performance Indicators (KPIs)	Quantifiable measurements, agreed to beforehand, that demonstrate the success of a project or an organizational unit within a company. They differ depending on the organization.

Microservice	A small, automated task that can be combined with other automated tasks and incorporated into larger services, enabling a modular approach to building large services/ solutions. eSignature is an example of a microservice.
Net Present Value (NPV)	The difference between the present value of cash inflows and the present value of cash outflows. NPV is used in capital budgeting to analyze the profitability of an investment or project. NPV accounts for the time value of money.
OpEx	Short for operating expenses. A category of expenditures that a business incurs as a result of performing its normal business operations. In IT, for example, this would include network, internet, email and other functions required to keep the business running. OpEx must be expensed each year and cannot be depreciated over time.
Platform as a Service (PaaS)	Delivery of a computing platform and solution stack as a service. PaaS offerings facilitate deployment of applications without the cost and complexity of buying and managing the underlying hardware, software and provisioning hosting capabilities. It provides all facilities required to support the complete lifecycle of building and delivering web applications and services that are made available on the internet.
Private Cloud	Computing environment (also called on-premise cloud) in which infrastructure is operated for a single organization. It can be managed and/or hosted internally or by a third party.
Product Manager (Service)	Individual responsible for defining, refining and ensuring that the overall stability, health and anticipated profit and loss benefits are obtained from a given service offering.
Profit and Loss Statement (P&L)	A financial statement that summarizes the revenues, costs and expenses incurred during a specific period of time, usually a fiscal quarter or year. These records provide information that shows the ability of a company to generate profit by increasing revenue and reducing costs. The P&L statement is also known as a statement of profit and loss, an income statement or an income and expense statement.

Public Cloud	Collection of external cloud computing resources (also called off-premise cloud) that are dynamically provisioned by offsite third-party providers to the general public on a self-service basis over the internet via web applications/ web services.
Scrum	Scrum is an iterative and incremental agile software development framework for managing product development.
Service Level Agreements (SLAs)	Part of a service contract in which a service is formally defined. The term SLA is sometimes used to refer to the contracted delivery time of the service or performance.
Software as a Service (SaaS)	Software delivery model in which an application and its associated data are hosted centrally (typically in the cloud) and accessed over the internet, usually from a thin client with a web browser.
Time to Value	The amount of time required to reach initial goals of either savings or increased revenue. Anticipated time to value is often used to create the business case for a given service or project.
Virtual Private Cloud (VPC)	An on-demand configurable pool of shared computing resources allocated within a public cloud environment, providing a certain level of separation and secure access among the different organizations using the resources.

APPENDIX A

The Business Requirements Document

IT professionals have differing opinions when it comes to Agile versus waterfall development. If you're building or equipping a data center, waterfall is a must. That's because a large initiative affects everything from the building itself to the wiring. Consequently, understanding the details is essential. Most software solutions, however, can be developed using Agile methodologies—as long as you apply the methodologies correctly.

Companies that struggle to implement Agile typically do so because they don't have a common blueprint when a solution/service is first defined. As a result, they don't have a guide for managing resource workstreams moving at different speeds and ensuring that the solution/service meets the minimum requirements of the business case.

While there's a big push to move to Agile development, companies need a combination of development methodologies to manage the complex world of IT while meeting the demands of the digital economy. The solution is a common blueprint that can be used across various development methodologies such as waterfall, iterative agile or Kanban agile.

A business requirements document (BRD) is similar to an architectural blueprint that depicts the overall structure requirements of a new building or a remodel. The blueprint enables homeowners, builders, architects and others to estimate costs, obtain approvals from local planning agencies and guide the construction activities. A BRD documents the basic structure and requirements of a solution or service. It provides a common language across business and technology teams, thereby helping to drive the success of cloud projects. It enables the owner to get management approvals, estimate costs and guide the development. In traditional waterfall development shops, BRDs are often hundreds of pages in length. In Agile shops, they are considerably shorter—from one to 25 pages.

Just as you don't need a blueprint for every home project, you don't need a BRD for every development project. Replacing kitchen cabinets or putting in new light fixtures won't affect the structure of the house, so you don't need to go to the expense of having

Figure 30. **Development methodology**

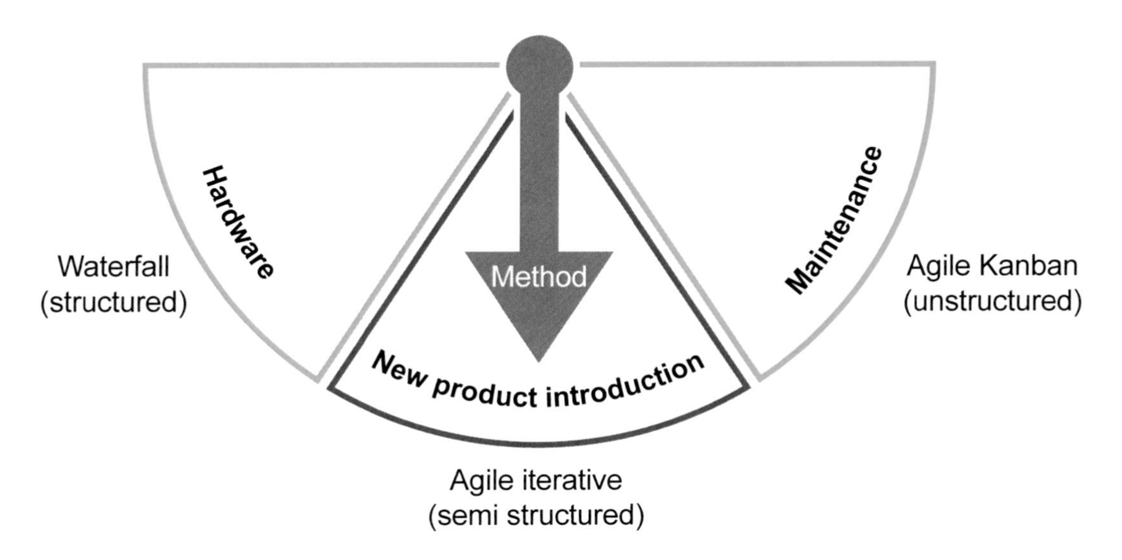

blueprints drawn up. Likewise, changing a link on a website or updating web content doesn't structurally affect a solution or service, so you don't need a BRD for those types of changes.

As Figure 30 shows, the more structured the methodology, the longer the requirements document must be. Kanban Agile, which is used for maintenance, is at the far end of the spectrum with respect to structure. The unstructured nature of Kanban increases the velocity of minor cosmetic changes. If a BRD is necessary at all for small incremental enhancements, it is very short.

Be aware that even if Agile is the primary development methodology, some projects require the more structured approach that the BRD provides. For example, setting up a self-service portal may require additional information such as details on the benefits, the users, SLAs and regulatory requirements. A BRD would be needed to ensure all stakeholders understand the business requirements, benefits and costs.

APPENDIX B

Getting Started on a Security Strategy

by Don Cox, Recovery.Gov

When it comes to cybersecurity, there is no one-size-fits-all approach that works for every company. Companies need to approach their strategies for protecting data and systems based on the areas of the business that generate revenue and the regulatory mandates and industry standards such as the Basel Accords, the Health Insurance Portability and Accountability Act of 1996 (HIPAA), the Sarbanes-Oxley (SOX) Act of 2002 and the Payment Card Industry Data Security Standard (PCI DSS).

So where do you begin? Start by asking if the company will insource or outsource its security needs. This question generates two additional questions: What will it cost and what is my risk? To get the answers, you must first inventory your people, technology and data resources and assess each one.

When it comes to people, ask if your budget allows you to hire someone to manage edge router, firewall, VPN, network, wireless, anti-virus and anti-malware, DMZ appliance, virtualization, storage (may include cloud), specialized applications, mobile device management, disaster recovery; provide end-user support, patch management, maintain compliance to company policy and legislative mandates; and enable IT to assess security parameters for employees who work at home.

Security costs and risks need to be managed wisely. Understanding what to insource and what to outsource requires an honest assessment of the talent pool in your organization. Your assessment should include both the executive leaders who manage security professionals and staff members who maintain security protocols and services. As a way to reduce costs and complexity, consider keeping critical security services in house and outsourcing the ones that do not require full-time attention. After the assessment is complete, estimate the cost of a full-time employee to manage the environment. Compare that cost with outsourcing the job, in which case you are essentially paying for a percentage of someone's time. Also consider outsourcing some of the technology services that aren't within the company's core competency. Examples include single sign-on, backoffice functions and cloud for public-facing data that isn't sensitive or confidential.

Invest in an independent review of your assessment. The best approach here is to negotiate with a company that will potentially be your partner for implementation and maintenance. Another possibility is a part-time CISO. Retirees and experienced CISOs are offering their skills and time, which might mean you don't have to budget for a full-time employee.

Be sure to consider physical as well as logical security. Data centers, offices and bring-your-own-device/choose-your-own-device strategies open up points of entry to your data. Hackers are now focusing less on the company itself to obtain secrets and more on the entities that are doing business with and have access to data for the company they are targeting.

A big component of your security plan is understanding the motivation of hackers—both external and internal. So ask yourself what the company has that someone might want. Once you identify what assets need protection, document comprehensive security plans, procedures and requirements. This includes documenting your environment, inventory assets and architectural maps. It also includes maintaining weekly and/or monthly status reports for virus detection, patch management and user account audit. For the latter, ensure you examine not only employees but also external providers who have admin privileges, checking for:

A big component of your security plan is understanding the motivation of hackers—both external and internal.

- People whose privileges were escalated
- People who always log in with the admin user account instead of their own
- How privileges are granted

The job isn't done once you've established your security policy. People have a tendency to think that once policies are in place security will manage itself and the people who are being paid to handle security are doing their jobs. In reality, complying with policies is often more difficult than developing them.

Hacker penetration testing (or pen testing) conducted at least once a year is a must. Your strategy should include steps for annually documenting and reviewing action plans for breaches. Monthly and new-hire training should be part of the plan along with scenario-based drills for IT and staff. The best plan is one that is thorough and practiced. Don't think in terms of what you will do if a breach occurs. Write an action plan that spells out how to protect critical assets and handle remediation when a breach occurs.

Here are some practical steps to protect your business:

- Implement acceptable technology use policies and have employees, staff and contractors sign off on them. Although documentation may not prevent behavior that could cause a breach, it does show due diligence and provides legal action that can be taken should one occur.
- Conduct regular and comprehensive security training for all employees and contractors. Include every level of the company from the cleaning crew to senior leaders. New types of breaches happen every day, so everyone must understand how to protect the company and themselves.
- Secure the assets not the perimeter. Focus your highest security efforts on locking down and locking out access to critical assets. For example, add a third layer of permission to encrypt databases, requiring manual signoff by the owner of the data as well as an IT person.

> Focus your highest security efforts on locking down and locking out access to critical assets.

- Practice responses to breaches. Keep up with the types of attacks that are occurring and ask what you would do if faced with similar attacks. How would your organization respond if a critical system was locked due to crypto locker, if someone discovered that files are missing or you learn that an employee possesses or is accessing material without authorization?
- Address physical building security. Make sure you have locks on doors into areas with hardware, systems and data. Maintain an up-to-date inventory and account for everything that can hold media. Take steps to ensure that your data cannot physically leave your premises. Ensure all password lists or accounts are not physically visible. Remember social engineering is a big risk to security.
- Implement security measures and recommendations from service providers. Cloud service providers and other vendors can provide information on best practices and security measures that are required for their systems to pass audit. Work with these partners to fine tune your security plans to ensure the entire environment is protected.

APPENDIX C

Architecting a Cloud Service Platform for IoT Hypergrowth

Industry experts predict exceptional growth for the Internet of Things (IoT) over the next four years. By 2020, more than 50 billion devices will be connected.[18] As Figure 31 shows, companies will connect everything from smart meters to cars to their networks via the cloud. Although many vendors

By 2020, more than 50 billion devices will be connected.

Figure 31. **Internet of Things**

18. "Internet of Things Lacks Safety Today, Opening Door to Major Threats Tomorrow, Warns OTA," Online Trust Alliance, August 11, 2015, https://otalliance.org/news-events/press-releases/internet-things-lacks-safety-today-opening-door-major-threats-tomorrow.

are embracing IoT from a strategic implementation perspective, only a few of the people interviewed for this book had considered the impact of IoT's dramatic growth as part of their overall hybrid cloud architectures.

Consider Security and Compliance

Many companies are enthusiastic about IoT but do not realize that the standards required for a secure and compliant solution are still lacking. Although resources are available to guide companies in creating a strategy and building a cloud platform, the standards still lack clarity and haven't been widely adopted across device manufacturers.

> The standards still lack clarity and haven't been widely adopted across device manufacturers.

Leading software vendors have expressed concern over the lack of safety and potential risks to interconnected systems and device. Several vendors have joined forces to establish the Online Trust Alliance (OTA) to facilitate the development of best practices and standards around IoT. OTA concerns include unsecured web interfaces, inefficient authentication, unsecured network services, lack of sufficient encryption, privacy, issues with mobile connectivity and limited ability to configure security into the devices.

As a result, CCoEs should consider the security and compliance risks of integrating a population of interconnected devices with legacy systems and applications because of IoT's current immaturity with respect to rules and controls. Limited oversight and the lack of best practices make introducing IoT into an enterprise environment a bit riskier: IoT devices are not as regulated as the rest of the systems on your network. iSpeak Cloud recommends that, in designing a cloud service platform, organizations should create a separate service category for IoT. Until standards and security mechanisms are in place, you'll need a different approach in the design of how the IoT service category interacts with the cloud service platform and other systems. Where possible apply best practices for security and standards from traditional standards bodies such as Cloud Security Alliance, National Institute of Standards and Technology and/or Distributed Management Task Force until the IOT standards have been solidified.

Compliance by Design

When building out the IoT connection to your cloud service platform, restrict access by APIs to protect data that the APIs don't need to access. Also consider adding layers of access. This approach provides additional protection at the design layer to reduce the possibility that ThingBots or other security threats will negatively affect your environment. The IoT category should address an additional layer of questions for connectivity to the cloud service platform and the security microservice. Questions to consider include:

- Is there a cost/benefit related to connecting the IoT devices for this category to the cloud service platform or should IoT devices be completely isolated from the platform used for other business operations?

- Does the business case for IoT include items that affect the cloud service platform and infrastructure such as scale, data management and higher ISP costs?
- Can the IoT component be upgraded? What type of security or mechanisms are there to turn off this capability and turn it on only as needed to avoid elevated privileges or a potential security breach?
- Who is involved in the design, coding and support? Include internal company employees as well as device manufacturers.
- How is security built into the device? Was it an afterthought or part of the initial design?
- How do these devices interact with existing services? Do they need to interact at all?
- Are there common default passwords? Are there passwords at all? How frequently must passwords be changed?
- Is there an excessive number of services running outside the devices' minimum requirements? If so, can they be disabled?
- Does the cloud service for this device take security seriously? Does the provider offer customers published security settings, prescriptive guidance and verifications?
- Does the device support a big data initiative for the company? If so, do these devices provide the data while ensuring integrity and confidentiality?
- What is the plan for device support after it connects to your network?

Build for Scale

A comprehensive BRD is a must for IoT implementations. The BRD should include forecasted growth, utilization and requirements for devices on your infrastructure as part of your cloud strategy. Although the data load may seem small at first, as more devices are brought into the environment and the number of data points collected per device grows, IoT will tax your infrastructure.

> A comprehensive BRD is a must for IoT implementations.

The chargeback/showback components of the cloud service platform should tie IoT costs to service owners for utilization of network, connectivity, storage and other resources. If there is a big data initiative, chances are the IoT strategy will be incorporated into it and both will drive up costs for network, storage and compute. This is because small data comes from devices that feed big data initiatives. Scaling involves more than just connecting devices.

Additional devices increase the demand for storage, reporting and other resources, so include these costs in the BRD and incorporate them into your design. Typically, because devices do not have a human element to consider in rollout, onboarding happens very quickly once the executive team realizes the return on investment for the company. Make sure your infrastructure is robust enough to accommodate the increased consumption of critical resources from end users and now devices.

APPENDIX D

Architect Cloud Strategy around Standards and Compliance

iSpeak Cloud with contributions from Mark Bodman, IT4IT Strategist, HPE

Architecting your cloud strategy with compliance in mind is vital to success. Yet planning your strategy around industry standards and best practices requires a shift from traditional IT thinking. Too many parallel shifts may create confusion and lead to internal conflicts. Consequently, it's wise to give some thought to the best way to leverage current industry and technology standards.

The fundamental iSpeak Cloud best practices, templates and other resources are derived from publications produced by widely accepted standards bodies, including IT Infrastructure Library, National Institute of Standards and Technology, International Standards Organization (ISO), Distributed Management Task Force, IT Process Institute and Cloud Security Alliance. Since ITPI published its *Visible Ops Private Cloud*, which is based on NIST and ITIL, newer standards organizations such as IT4IT, International Association of IT Asset Managers (training/cert) and DevOps have developed prescriptive guidance that complements the work done by the organizations just mentioned.

Properly integrating cloud capabilities into the fabric of an IT organization requires an evolutionary shift in the IT operating model. Most IT organizations today are structured to deliver projects instead of services. Regardless of the model you adopt, you must follow the core principles and align them across the company. One *Visible Ops Private Cloud* core principle is the need to shift from a project or resource view to a service view for strategic cloud initiatives to be successful. Companies must become skillful at planning, developing and sourcing, and providing customer access to a portfolio of services. They must then fulfill and manage those services effectively through their lifecycle. An important iSpeak Cloud core principle is the requirement to go beyond the service broker role, evolving IT into a role of a facilitator and partner in driving technology strategies.

Mapping Cloud Strategy to Business Needs

iSpeak Cloud recommends that you map your cloud strategy to the overarching company standards and process initiatives to ensure everyone is on the same page. This enables you to resolve conflicts across standards and reduce political churn, thereby speeding digital transformation. Competing standards and processes often lead to conflicts, missed deadlines and resource contention. To avoid these obstacles, your process engineers must rethink their approach by evaluating the various standards and best practices and selecting the ones that work best for your company.

You will need to map existing and emerging standard adoption with legacy policies, processes and standards. For example, your company may already be leveraging best practices from *Visible Ops Private Cloud*, ITIL, NIST and DMTF to move to a private cloud and establish a service portfolio view as part of service management. Perhaps you have decided to leverage a standard such as IT4IT from the OpenGroup to drive that portfolio strategy within IT to a more detailed level to broker more essential IT services. At the same time, as the next step in the evolution of maturity, IT is trying to transform itself from a service broker to a trusted advisor or partner.

A unified approach to transformation, as shown in Figure 32, is the essential ingredient to success. Such an approach includes mapping out unifying process overlap, identifying guidance applicability and overlap, and following guidance that makes sense for your company's goals and objectives.

Figure 32. **Steps to transformation**

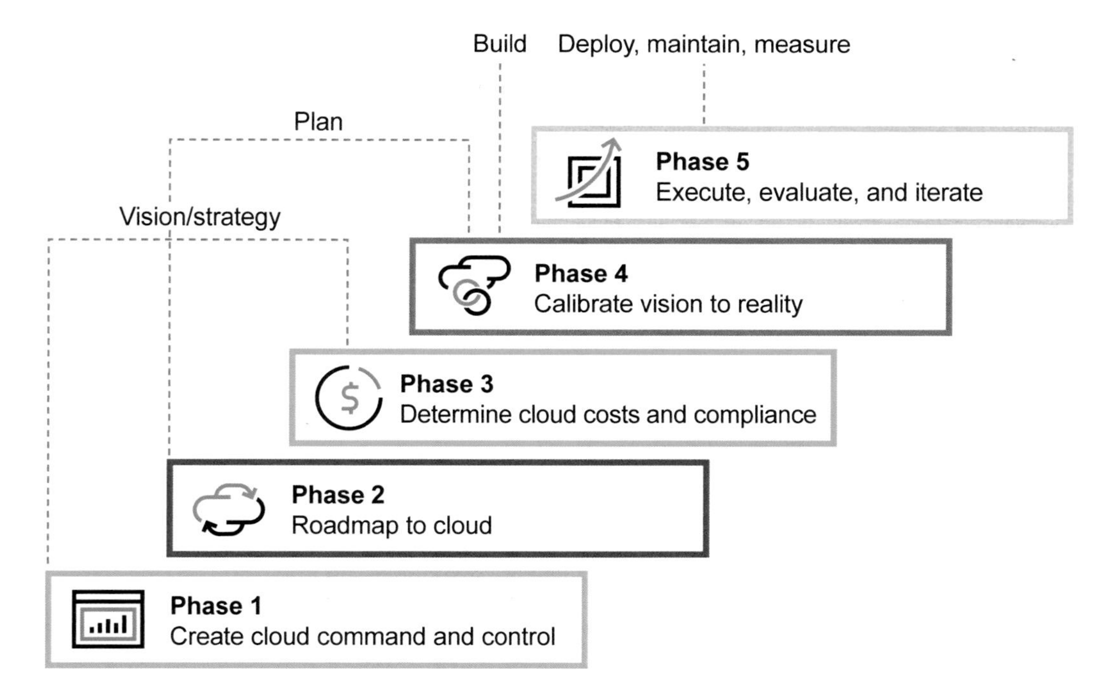

In the case of IT4IT and iSpeak Cloud, it is service lifecycle management. The company has identified four phases needed for successful digital transformation across the business and IT for service lifecycle management. Those phases are:

- Vision/strategy for the company
- Service plan
- Building the service
- Continuous deployment, maintenance and measurement

The iSpeak Cloud phases may encompass many areas of the service lifecycle and vice versa. For example, vision/strategy can be covered by steps such as creating a cloud center of excellence, high-level roadmap and initial costing for the company business case. However, standards such as IT4IT may not map back to the service lifecycle vision/strategy phase because they start after it has been approved and budgets allocated from a higher level and initiatives pushed back into IT. You may also find that a service lifecycle phase may encompass multiple steps within a standard. In the Figure 33, two of the four value streams for IT4IT are covered under the deploy, maintain and measure phase of the service lifecycle.

Figure 33. **The four value streams map back to the steps to transformation phases in Figure 32**

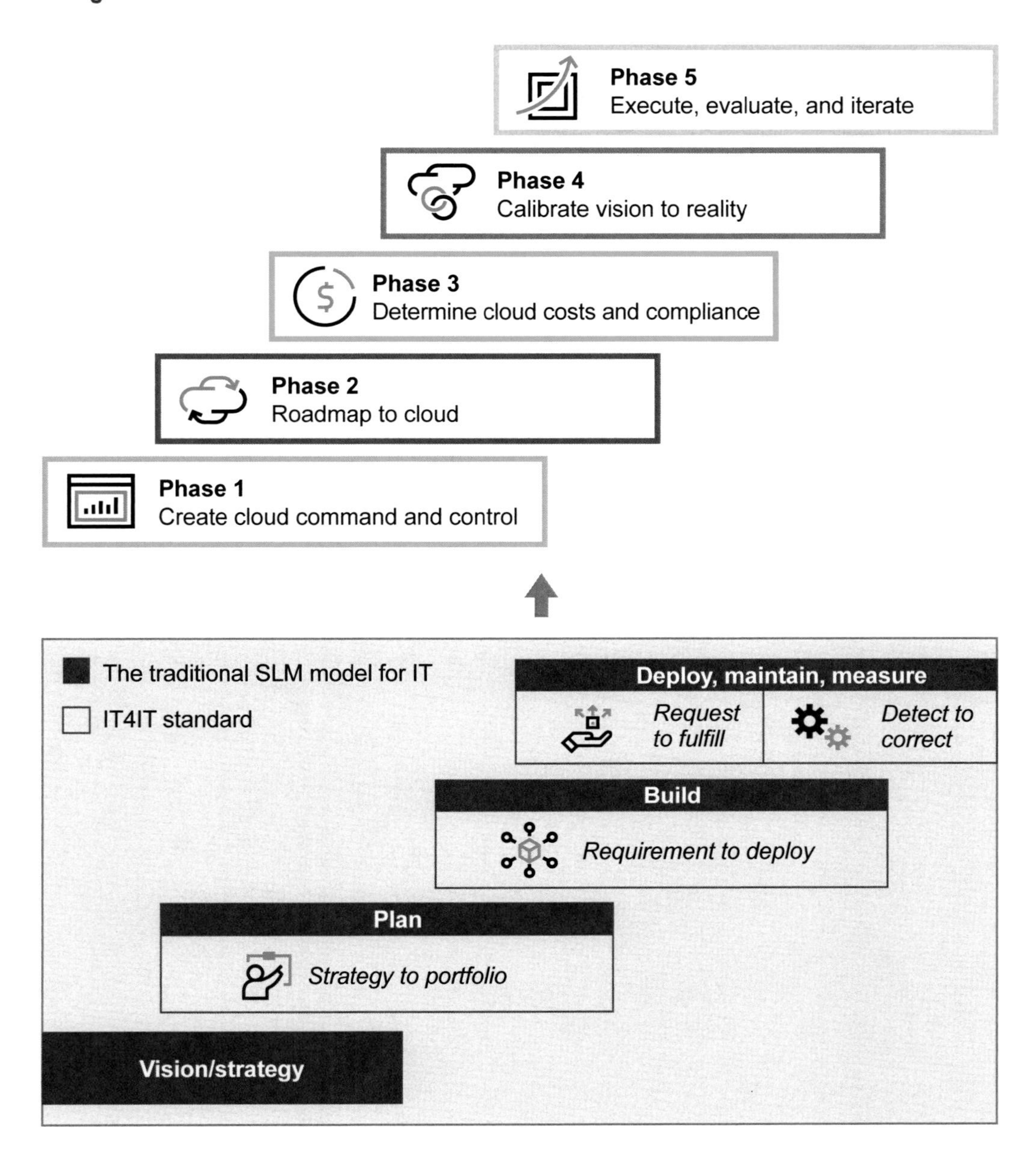

The service lifecycle management model and IT4IT standard overlap the phases, as you can see by the way they align.

New Role: Process Strategist

Just as you would have the cloud service delivery manager and data strategist translate requirements for their functional areas, you may want to create a new role of process strategist. This role can help identify and resolve potential areas of conflict among the various standards and best practices (existing and emerging) in the market. The process strategist can help merge critical areas such as ITIL and IT4IT along with your CCoE framework to avoid conflict. For example, the cloud service platform encompasses the microservices/capabilities foundation and management fabric to manage workloads across clouds and automate many of the legacy processes around incident, problem, change and configuration management. Perhaps the strategist will be an expert in closed-loop incident process (CLIP). Through his or her expertise in CLIP, the strategist can provide prescriptive guidance on how a process currently works, identify requirements for automation and negotiate with the process owner to define what the new process will look like. To learn more about the referenced standards and mapping them to one another, reach out to the following organizations for more information.

IT4IT	Open Group	www.opengroup.org
DMTF	Distributed Management Task Force	www.dmtf.org
NIST	National Institute of Standards and Technology	www.nist.gov
CSA	Cloud Security Alliance	https://cloudsecurityalliance.org
ISACA	Formerly Information Systems Audit and Control Association	www.isaca.org
OTA	Online Trust Alliance	https://otalliance.org
ITPI/Visible Ops	IT Process Institute	www.itpi.org